SpringerBriefs in Computer Science

For further volumes:
http://www.springer.com/series/10028

Heming Wen · Prabhat Kumar Tiwary
Tho Le-Ngoc

Wireless Virtualization

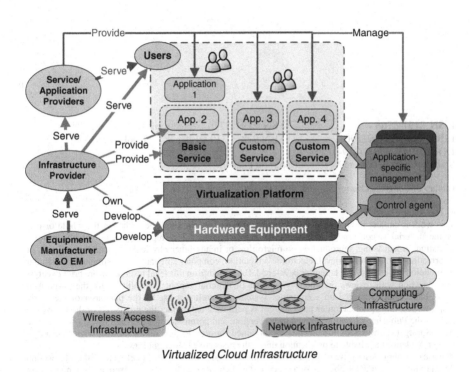

Virtualized Cloud Infrastructure

Springer

Heming Wen
Prabhat Kumar Tiwary
Tho Le-Ngoc
Department of Electrical and Computer Engineering
McGill University
Montreal, QC
Canada

ISSN 2191-5768   ISSN 2191-5776   (electronic)
ISBN 978-3-319-01290-2   ISBN 978-3-319-01291-9   (eBook)
DOI 10.1007/978-3-319-01291-9
Springer Cham Heidelberg New York Dordrecht London

Library of Congress Control Number: 2013943568

Printed on acid-free paper

Springer is part of Springer Science+Business Media (www.springer.com)

Acknowledgments

This work was partially supported by the Natural Sciences and Engineering Research Council (NSERC) through a NSERC Discovery Grant and the NSERC Strategic Network for Smart Applications on Virtual Infrastructure (SAVI), and the Fonds québécois de la recherche sur la nature et les technologies (FQRNT) via a scholarship.

Acknowledgments

Contents

Acronyms

3GPP	3rd Generation Partnership Project
ADC	Analog to digital converter
AKARI	AKARI (a small light) Architecture Design Project
AM	Allocation and management
AP	Access point (wireless)
API	Application programming interface
APSD	Automatic power save delivery (802.11)
ASN	Access service network (WiMAX)
BBU	Baseband processing unit
BPEL	Business Processes Execution Language
BS	Basestation
BSS	Basic service set (802.11)
BSSID	Basic service set identifier (802.11)
C&M	Control and management
CAPWAP	Control and provisioning of wireless access points
CDMA	Code division multiple access
CoMP	Coordinated (or cooperative/competitive) multi-point
CPU	Central processing unit
CSN	Connectivity service network (WiMAX)
DCF	Distributed coordination function (802.11)
DMA	Direct memory access
DPI	Deep packet inspection
DPID	Datapath identifier
DSA	Dynamic spectrum allocation/access
DSP	Digital signal processing
DTLS	Datagram Transport Layer Security
eNB	Evolved Node B (or E-UTRAN Node B)
E-UTRAN	Evolved Universal Terrestrial Radio Access Network
EDCA	Enhanced distributed channel access (802.11)
EPC	Evolved Packet Core
ESSID	Extended service set identifier (802.11)
FDMA	Frequency-division multiple access
FEC	Forward error correction

FFT	Fast Fourier transform
FIRE	Future Internet Research and Experimentation
FP7	Seventh Framework Programme of the European Union
FPGA	Field-programmable gate array
FSM	Finite-state machine
GC	Guest controller
GENI	Global Environment for Network Innovations
GPP	General purpose processor
GRE	Generic routing encapsulation
GWCN	Gateway core network (LTE)
HARQ	Hybrid automatic request query
HCCA	HCF controlled channel access (802.11)
HCF	Hybrid coordination function (802.11)
HDL	Hardware description language
HTTP	Hypertext Transfer Protocol
I/O	Input/output
IaaS	Infrastructure-as-a-service
IEEE	Institute of Electrical and Electronics Engineers
IETF	Internet Engineering Task Force
IOMMU	I/O memory management unit
IOVM	I/O virtualization manager
JGN-X	Japan Gigabit Network
JSON	Java Script Object Notation
LPFC	Linear proportional feedback control
LTE	Long Term Evolution
LUT	Look-up table
LVA	Local virtualization agent
LWAPP	Lightweight access point protocol
KVM	Kernel-based virtual machine
M2M	Machine-to-machine (communications)
MAC	Medium access control
MIB	Management information base (SNMP)
MIMO	Multi-input multi-output
MME	Mobility management entity (LTE)
MNO	Mobile network operator
MPG	Mobile personal grid
MPLS	Multiprotocol label switching
MU-MIMO	Multi-user MIMO
MVNO	Mobile virtual network operator
NAT	Network address translation
NETCONF	Network configuration protocol
NGN	Next-generation network
NIC	Network interface card
NVS	Network virtualization substrate
OF	OpenFlow

OFDMA	Orthogonal frequency-division multiple access
OFELIA	OpenFlow in Europe: Linking Infrastructure and Applications
ONF	Open Networking Foundation
ORBIT	Open-Access Research Testbed for Next-Generation Wireless Networks
OS	Operating system
OVS	Open vSwitch
PAN	Personal area networking
PCF	Point coordination function (802.11)
PCI	Peripheral Component Interconnect
PCIe	Peripheral Component Interconnect Express
PLC	PlanetLab Central
PRB	Physical radio block (LTE)
PSM	Power saving mode (802.11)
QEMU	Quick Emulator
QoS	Quality-of-service
RAC	Radio access control
RCB	Radio control board
RFC	Request for comment (IETF)
RPC	Remote procedure call
RTT	Round-trip time
SAE	System Architecture Evolution
SAVI	Smart Application on Virtual Infrastructure
SDMA	Space-division multiple access
SDN	Software-defined networking
SDR	Software-defined radio
SIE	Slice isolation engine
SIMD	Single instruction multiple data
SLA	Service-level agreement
SLN	Service with leased network
SNMP	Simple network management protocol
SOA	Service-oriented architecture
SQL	Structured query language
SR-IOV	Single-root input/output virtualization
SSH	Secure shell
SSID	Service set identifier (802.11)
STA	Station (802.11)
TDMA	Time-division multiple access
UDP	User Datagram Protocol
UE	User equipment
USRP	Universal Software Radio Peripheral
VANI	Virtualized Application Networking Infrastructure
VAP	Virtual access point
vBTS	Virtual base transceiver station
VLAN	Virtual local area network

VM	Virtual machine
VMI	Virtualization manager interface
VMM	Virtual machine monitor
VN	Virtual network
VNC	Virtual network controller
VNP	Virtual Node Project
VNTS	Virtual network traffic shaper
VPN	Virtual private network
VPR	Virtualized physical resources
VRM	Virtual resource manager
WARP	Wireless Open-Access Research Platform
WLAN	Wireless local area network
WMAN	Wireless metropolitan area network
WMM	Wi-Fi multimedia
WSDL	Web Service Description Language
WTP	Wireless termination point
XML	Extensible markup language
XMPP	Extensible Messaging and Presence Protocol

Custom Acronyms for Multi-Dimensional Framework

CMM	Custom management module
CML	Control and management layer
GIM	Gateway interface module
IAM	Interface abstraction module
IDC	Inter-domain communication
LVA	Local virtualization agent
LVAC	Local virtualization agent contact
M-mux	Management multiplexer
M-visor	Management hypervisor
M&M	Monitoring and measurement
MMA	Monitoring and measurement agent
MFC	Management function client
MFS	Management function server
MTT	Management tools and templates
SRM	Spectrum reshaping module
VL	Virtualization layer
vWPL	Virtual wireless protocol layer

Chapter 1
Introduction

1.1 Virtualization as a Concept

Hardware virtualization is a well-applied concept, particularly in the field of computing technologies. It involves the *abstraction* and *sharing* of hardware resources among different parties. Virtual memory systems and virtual machines are features present in modern computers and operating systems that are often referred to as *server* or *host virtualization*. With the ever-increasing importance of virtualization-derived services such as cloud computing, virtualization has proven itself to be a key component in the evolution of computing technologies. Naturally, virtualization can be extended from a single host to a network of hosts. The so-called *network virtualization* (a more comprehensive definition will be given in Sect. 2.3) is an active area of research that attempts to bring virtualization to the wired network fabric. However, the quest for virtualization does not end here. In recent years, the constant growth of wireless access technologies highlights its potential as a powerful extension to the wired infrastructure. Thus, it is undeniable that the future of digital communications will heavily rely on wireless technologies. As the virtualization trend progresses, the decision to bring virtualization to the wireless access is now being considered.

Just like computing and networking resources, wireless resources are also shared among many users. Nevertheless, as opposed to the relatively stable and controllable wired medium, the wireless medium is an ever-changing mobile environment. As a result, there is a need for the virtualization techniques used in host and network virtualization to be modified or adapted when applied to wireless technologies. The purpose of this book is to present the background knowledge required for the readers to understand the concept of wireless virtualization, its relationships with other technological trends and its potential challenges. In order to provide a context for wireless virtualization, some network virtualization architectures and technologies are also covered. Thus, this book contains an overview of virtualized infrastructures both *with* and *without* wireless resources. This book will hopefully facilitate the formulation of a sustainable wireless virtualization framework.

H. Wen et al., *Wireless Virtualization*, SpringerBriefs in Computer Science,
DOI: 10.1007/978-3-319-01291-9_1, © The Author(s) 2013

Of course, although the coverage of this book is broad, it is in no way a comprehensive or authoritative guide to wireless virtualization. To complement the reading of this book, other survey documents will be pointed out as references. For instance, an overview of emerging wireless technologies that are discussed in this book as either *enablers* or *applications* of wireless virtualization can be found in [1].

1.2 Motivation for Wireless Virtualization

Prior to network virtualization, the performances of applications and protocols were tied to the fixed configuration of the network hardware, causing operational inflexibility. To maintain a truly-differentiated service on the underlying network infrastructure was often impossible due to the lack of flexibility in the network architecture. To amend the situation, alternative *clean-slate design* is used to derive the next generation of network [2]. One question that arises is the role that wireless virtualization can play, if any, in the future network infrastructure. According to a white paper jointly published by major network providers across the world in [3], the decoupling of the infrastructure from its functionalities is one major step in the future of networking technologies and is referred as *network functions virtualization*. One of the important network functions mentioned in [3] is the *mobile network node*.

As discussed in the introductory section, wireless technologies have gained considerable momentum due to the explosive growth of mobile computing platforms (i.e. smart phones). However, the constant influx of new and amended wireless standards has created a rich but chaotic wireless environment in which multiple standards are competing and coexisting. In such heterogeneous environment, the interoperability and the resource allocation among different technologies are some of the potential issues to solve. The future is all about the *coexistence* and *convergence* of wireless technologies in a *service-oriented infrastructure*. The abstraction provided through wireless virtualization is one possible solution to both simplify and unify existing heterogeneous networks. As such, it can enable smooth and ubiquitous transition among different technologies [4]. Overall, the implementation of wireless virtualization includes benefits in both industrial and academic contexts.

From a commercial perspective, by enabling a flexible sharing and reuse of the existing infrastructure, virtualization can lower the capital expenditures and the barrier to entry for emergent mobile service providers [5]. New emerging technologies such as *machine-to-machine* (M2M) communications that require highly-customized capabilities can be integrated and deployed over the existing infrastructure. At the same time, various network functions can be consolidated into datacenters, leveraging from their high-volume processing capabilities and highly flexible provisioning [3]. This relocation of functionalities away from the physical hardware can be supported by the advent of low-latency optical paths. Additionally, virtualization effectively allows the wireless hardware infrastructure to be provided as a *service* instead of a physical asset. Combined with *software-defined networking* (SDN), virtualization allows the introduction of a *service-oriented architecture* (SOA) in wireless network infrastructures. This integrates the wireless access network into an *extended cloud infrastructure*. The virtualized infrastructure can then be expanded and provisioned *on demand*. The ability for the service provider to control and customize the underlying infrastructure without the need to own it leads to more efficient operations and better quality of service (QoS). In other words, a more integrated, immersive, diverse and competitive mobile environment can be offered to the rapidly-growing market of mobile users [4]. Wireless virtualization is one possible enabling technology for a *converged multi-service* infrastructure.

In terms of research applications, large experimentation projects for the development of Future Internet and next-generation networks (NGNs) require a level of flexibility and sharing that can be achieved through virtualization. Thus, research projects following *clean-slate design* extensively use virtualization techniques [2]. Since any future network infrastructure will have ubiquitous integration of wireless access technologies, wireless virtualization must be considered. The development cycle for new wireless technologies can be shortened if a more open and flexible wireless infrastructure is available [7]. Overall, on-demand deployment of new services and applications through virtualization will accelerate the development of a more innovative and dynamic wireless environment in both commercial and academic context.

1.3 Structure of the Book

This book is divided into chapters, sections and subsections. The first part of the book covers network virtualization in order to provide a context for wireless virtualization. However, the contents of each chapter and section are not necessarily mutually exclusive. Many ongoing research projects span across multiple layers of the communication system and are applicable to both network and wireless virtualization. The structure of this book has the goal to provide a more comprehensive picture on the role of wireless virtualization in a virtualized infrastructure. This book attempts to address concerns such as the application of existing network virtualization concepts on wireless resources and the apparent disparity among different wireless virtualization approaches.

In Chap. 2, existing surveys on network virtualization will be summarized and compared. One possible classification scheme extracted from [6] will be briefly discussed. In Chap. 3, meta-architectures for large-scale virtualization testbeds such as the Global Environment for Network Innovations (GENI), the Smart Application on Virtual Infrastructure (SAVI) and the project AKARI will be reviewed. Technologies and techniques that can be applied in network virtualization, such as OpenFlow and virtual network interface cards (NIC), will be examined in Chap. 4. This will be followed by an overview of the different aspects of wireless virtualization in Chap. 5. Multiple access techniques, management frameworks and example applications will be respectively presented. The virtualization of 802.11 wireless local area networks (WLAN) and cellular basestations (WiMAX and LTE) will be discussed. The application of software-defined radio (SDR), spectrum slicing and RF frontend slicing in virtualization will also be covered in Chap. 5. Finally, Chap. 6 will present a comparative summary of the various virtualization architectures and technologies surveyed in this book. The last chapter is also dedicated to provide a unified view on the different wireless virtualization perspectives. It contains a prototype of a sustainable multi-dimensional virtualization framework for wireless access networks.

References

1. D. Raychaudhuri, M. Gerla (eds.), *Emerging Wireless Technologies and the Future Mobile Internet* (Cambridge University Press, New York, 2011)
2. S. Paul, J. Pan, R. Jain, Architectures for the future networks and the next generation Internet: A survey. Comput. Commun. **34**(1), 2–42 (2011)
3. Network functions virtualization: An introduction, benefits, enablers, challenges and call for action. Introductory white paper, SDN and OpenFlow world congress, Oct 2012. Available: http://www.tid.es/es/Documents/NFV_White_PaperV2.pdf
4. K.K. Yap, R. Sherwood, M. Kobayashi, T.-Y. Huang, M. Chan, N. Handigol, N. McKeown, G. Parulkar, Blueprint for Introducing innovation into wireless mobile networks, in *Proceedings of the Second ACM SIGCOMM Workshop on Virtualized Infrastructure Systems and Architectures*, Sep 2010
5. Y. Zaki, L. Zhao, C. Görg, A. Timm-Giel, A novel LTE wireless virtualization framework, in *Proceedings of the Second International ICST Conference on Mobile Networks and Management*, Sep 2010
6. A. Wang, M. Iyer, R. Dutta, G. Rouskas, I. Baldine, Network virtualization: technologies, perspectives, and frontiers. J. Lightwave Technol. **31**(4), 523–537 (2012)
7. J. Sachs, S. Baucke, Virtual radio: a framework for configurable radio networks, in *Proceedings of the 4th Annual International Conference on Wireless Internet*, Nov 2008

Chapter 2
Network Virtualization: Overview

Before presenting different network and wireless virtualization architectures and techniques, this chapter summarizes what other existing surveys on similar topics have already covered. However, the readers might wonder *why network virtualization is discussed in great details in a book on wireless virtualization?* In fact, the so-called *wireless virtualization* borrows many concepts and benefits from network virtualization. Additionally, wireless virtualization must *coexist* within the *context* of network virtualization in most applications. One of the goals of the first half of this book, (this chapter and Chaps. 3, 4) is to highlight different concepts and implementation technologies used in network virtualization that can be relevant, applied, integrated or extended to wireless virtualization. In this chapter, the readers are introduced to *virtualization*, a popular but often misused term. First, the evolution from computer and server virtualization to network virtualization is briefly discussed. Then, a classification tree to organize different virtualization architectures and technologies is constructed based on existing surveys. At last, an interpretation of the definition of *network virtualization* is given. The clear link between the network virtualization concepts presented in this chapter and wireless virtualization is emphasized.

2.1 From Server Virtualization to Network Virtualization

As mentioned in the previous chapter, virtualization technologies have recently moved from server virtualization to network virtualization. *Server virtualization*, also known as *host* or *computer virtualization*, allows multiple users to *share* the same server through *virtual machines* (VMs) by abstracting and decoupling the computing functionalities from the underlying hardware. Server virtualization has played a pivotal role in cloud computing as one of its main enablers. Through server virtualization, *on-demand provisioning* and *flexible management* of computing resources are made possible. Strictly speaking, server virtualization also includes the virtualization of network interfaces from the operating system point of view. However, it does not involve any virtualization of the network fabric, such

H. Wen et al., *Wireless Virtualization*, SpringerBriefs in Computer Science, DOI: 10.1007/978-3-319-01291-9_2, © The Author(s) 2013

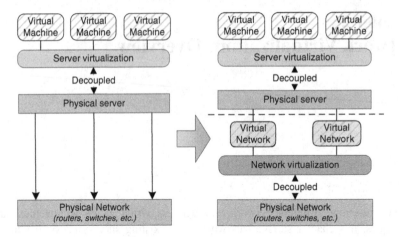

Fig. 2.1 Server virtualization versus network virtualization

as switches and routers. On the other hand, *network virtualization* enables multiple isolated virtual networks to share the same physical network infrastructure, as shown in Fig. 2.1. Thus, it decouples the network infrastructure from the services that it provides. This paradigm shift allows virtual networks with truly differentiated services to coexist on the same infrastructure, maximizing its reusability. These virtual networks can be deployed on demand and dynamically allocated, just as VMs in server virtualization would. The *functionalities* that each virtual network can obtain from the virtualized infrastructure range from simple *connectivity* and *performance guarantees* to advanced support for *new network protocols*.

Along with the recent development in software-defined networking (SDN) technologies, network virtualization has also become involved in cloud computing applications. For instance, network virtualization enables a more flexible management of the network interconnection between physical servers in a large datacenter. New application of network virtualization can eventually allow the cloud services to extend beyond the datacenter into the network infrastructure. Using similar logic, the interesting question that arises here is whether virtualization can now be carried beyond the core network infrastructure and into the *wireless access network*. In other words, the feasibility and applicability of virtualization concepts in wireless access technologies must be considered. As explained in the Chap. 1, there are compelling motivations to do so.

2.2 Network Virtualization: Diverse Perspectives

Some surveys on network virtualization can be found in literature such as [1] and [2]. The survey [1] summarizes basic network virtualization concepts and covers selected virtualization testbeds developed by different research communities.

The survey [2] is an extension of [1] by presenting additional commercial technologies and formulating its own unified definition for network virtualization. According to survey [2], the term *network virtualization* is a very general term that can refer to *different concepts* under *different contexts*. It is often associated with resource allocation, virtual private network (VPN), virtual local area network (VLAN), multi-user multiple access techniques, SDN and a number of different concepts and technologies. In simplified scenarios, network virtualization is thought to be a technique of *network resource sharing*. However, depending on the context, it could be a combination of *resource abstraction, resource partitioning, resource aggregation, centralization of control, separation of data plane and control plane* and much more. To fully understand network virtualization, different *aspects* of it must be considered. An attempt to reorganize the different architectures has been made in [2]. Figure 2.2 represents a reconstruction of that classification. Interestingly, *wireless virtualization* also encompasses such complexities and will be approached in a similar fashion.

In Fig. 2.2, a separation between the industrial perspective and the academic perspective is made as in [2]. The term *industrial* can be contested because many of these commercial technologies were initially developed through academic research. It is more appropriate to view the component on the left-hand side of Fig. 2.2 as more matured technologies and implementation techniques [1]. The network virtualization techniques have been classified based on the network component where the virtualization is implemented. This format of classification is very helpful at providing some insight into what specific resources are being virtualized in each case. The right branch of the tree represents testbed architectures and management protocols that can incorporate a combination of specific technologies. This "academic" perspective mainly consists of emerging technologies and active research testbeds. However, these technologies will eventually mature

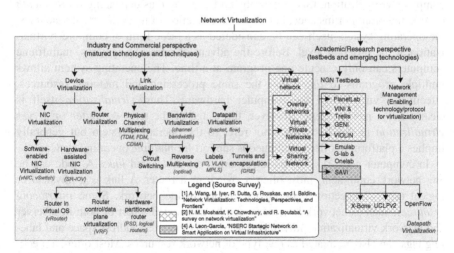

Fig. 2.2 Different aspects of network virtualization (based on [1], [2] and [4])

and merge with the commercial perspective. For instance, OpenFlow [3] will join *datapath virtualization* as a powerful alternative to label switching and tunneling. Many of these testbeds and technologies, including OpenFlow, will be respectively presented in Chaps. 3 and 4.

Once again, an important question that arises is: *where does wireless network virtualization fit in the context of network virtualization?* The survey [2] presents wireless virtualization as one of the *frontier* research areas. Many of the wireless network virtualization techniques covered in this book can fall into the category of *device virtualization*. This is because the wireless network, formed by devices such as wireless access points and cellular basestations, lies at the *edge* of the core network. In addition, *wireless network management* is an important aspect that often requires collaboration with core network management functions in a virtualized wireless infrastructure. Nevertheless, the classification for wireless network virtualization is *multi-dimensional*. Some aspects of it are unique to wireless technologies and remain independent from core network concepts. The different aspects of wireless virtualization will be discussed in Chap. 5. The different *wireless virtualization perspectives* will be derived in Chap. 6.

2.3 Definition of Network Virtualization

Discussion in [2] pointed out that the usage of the term *network virtualization* is highly overloaded. Thus, an attempt to unify the various definitions is made in [2]. In this book, the definition of *virtualization* in the context of *computer virtualization* is the *abstraction* and *sharing* of *computing resources* to give multiple parties the illusion of sole ownership [2]. In computing, the basic resources are processing power, active memory, data storage and driver interfaces. Nonetheless, even within computer virtualization, this definition can be ambiguous depending on the *depth* of virtualization. For instance, before the introduction of modern virtual machines, the operating system itself can be considered as a basic form of computer virtualization at the program level. Before the advent of operating systems, mainframe computers can only execute one program at a time. The operating system allows multiple programs to *time-share* the same processing and memory resources, effectively 'virtualizing' the computer hardware. Thus, *virtualization* itself is clearly not a new concept. However, the usage of the term *computer or server virtualization* in this book does not refer to the operating system but generally implies a platform to offer simultaneous support for *multiple* operating systems.

In computer networks, the basic resources are *nodes* and *links*. A node can be any network equipment, most often a switch or a router. A link is a physical or logical (soft) connection between two nodes in the network. Some *interfacing resources*, such as network interfaces, can be considered a part of both host/server and network virtualization. *Identification resources* such as address space and labeling tags can also be treated as one type of network resources. Most often, a mechanism for the management of the identification for devices (IP address and MAC address), links (flow ID), and networks (VLAN ID, VPN ID) is needed for the management and monitoring of a virtualized infrastructure. According to [2], identification resources are *soft resources* used to *assist* the virtualization of nodes and links. They are not considered to be *physical resources* on their own. High-level organizational structure such as a *network topology* can be created by assembling both soft and physical resources. Therefore, network virtualization is a *multi-dimensional* and *multi-layered* concept. This explains why the term *virtualization* is employed very loosely and often confusingly. After all, traditional VPN and VLAN can be considered a form of (primitive) network virtualization. However, many of the latest network virtualization projects can support more flexible functions beyond what VPN or VLAN can provide, such as the dynamic on-demand provisioning of resources and the support for entirely new network architectures.

The definition presented in the survey [2] covers very well the diverse spectrum of network virtualization by combining two basic concepts. Firstly, [2] clarifies the purpose of network virtualization, which is to provide a user or an application with an *illusion* of the *full ownership* of a network resource. Then, [2] describes the methods used to achieve this objective. In short summary, network virtualization can be defined as the *sharing of network resources* through the *abstraction* and *isolation* of *network functionalities* of the physical hardware. It is important not to confuse network virtualization with software-defined networking (SDN). SDN can be one of the *enabling technologies* for virtualization because by definition it provides abstraction and decoupling of functionalities from the hardware. However, the subtlety is that SDN by itself does not *necessarily* involve sharing and isolation.

This definition for network virtualization is broad enough to be applied to wireless virtualization to some extent. In the case of wireless virtualization, the purpose is to provide the mobile virtual network operators (MVNOs) with an abstracted ownership of the wireless access network. One can argue that wireless virtualization is simply a *subset* of network virtualization. One possible interpretation is that the wireless access points are equivalent to the network nodes and that the wireless channels are equivalent to the network links. However, the wireless medium is fundamentally different from the wired medium. This is due to the time-varying nature of the wireless channel and the high mobility of the end users. These wireless-specific issues will be addressed by different aspects of wireless virtualization and will be discussed in Chap. 5. Finally, the term *hypervisor* refers to the layer responsible for the isolation among slices of the virtualized infrastructure. It is often referred to as virtual machine monitor (VMM) in the context of computer virtualization.

References

1. N.M. Mosharaf, K. Chowdhury, R. Boutaba, A survey on network virtualization. J. Comp. Telecommun. Networking **54**(5), 862–876 (2010)
2. A. Wang, M. Iyer, R. Dutta, G. Rouskas, I. Baldine, Network virtualization: technologies, perspectives, and frontiers. J. Lightwave Technol. **31**(4), 523–537 (2012)
3. N. McKeown, T. Anderson, H. Balakrishnan, G. Parulkar, L. Perterson, J. Rexford, S. Shenker, J. Turner, OpenFlow: Enabling innovation in campus networks, in *ACM SIGCOMM Computer Communication Review*, Apr 2008
4. A. Leon-Garcia, NSERC strategic network on smart applications on virtual infrastructure. University of Toronto, Sep 2011. Available: http://www.savinetwork.ca/wp-content/uploads/Al-Leon-Garcia-SAVI-Introduction.pdf

Chapter 3
Network and Infrastructure Virtualization Architectures

This chapter covers virtualization applied in the context of *meta*-architectures and next-generation network (NGN) testbeds, which can also be referred to as *testbed virtualization*. In this chapter, the main focus is on the *organizational architecture* and the *management framework* of such architectures. It mainly describes the specifications of large testbeds from existing national and international research initiatives on Future Internet network architectures. Overall, this chapter provides a "big picture" and a context in which wireless virtualization can ultimately be integrated in. While many of these architectures do not explicitly address wireless virtualization, they do provide some universal design principles for a virtualized infrastructure. Additionally, the importance of wireless technologies in the Future Internet architecture is clearly recognized. Ultimately, a framework for wireless infrastructure virtualization can be derived based on the concepts summarized in this chapter.

The *testbed virtualization* architectures presented in this chapter are influenced by the "clean-slate" design principle, a main recurrent theme motivated by the need for a change in the increasingly-patchy Internet architecture. For instance, the inherent lack of security in the current IP architecture, the increasing demand for mobility and the shift from host-to-host applications to content-oriented applications prompt the development of new network paradigms [1]. Thus, one common objective of NGN testbeds is to *overhaul* the existing network architecture in order to support a *smarter* and more *open* infrastructure. A portion of this chapter is based on the survey articles [2] and [1] on current research trends in Internet technologies. The main projects of interest include the Global Environment for Network Innovations (GENI) in USA, the Smart Application on Virtual Infrastructure (SAVI) [3] in Canada and the AKARI Project in Japan. Specific *application* and *service-oriented* testbed architectures related to each of these national initiatives are respectively presented. These include PlanetLab, the Virtualized Application Networking Infrastructure (VANI) and the Virtual Node Project (VNP) from Japan Gigabit Network (JGN-X). The Future Internet Research and Experimentation (FIRE) project from the European Union is also featured in [2] but is only briefly mentioned in this chapter. On the other hand, derivative works based on the European initiative 4WARD related to wireless

H. Wen et al., *Wireless Virtualization*, SpringerBriefs in Computer Science,
DOI: 10.1007/978-3-319-01291-9_3, © The Author(s) 2013

virtualization will be discussed in Chap. 5. Overall, the overview of virtualization architectures presented in [1] and [2] are combined with information extracted from publications specific to each of the selected testbeds.

3.1 Global Environment for Network Innovations

GENI is currently one of the largest and most complex virtualization-enabled test-bed *meta*-architecture in development, initiated by the American National Science Foundation. It runs on the dedicated backbone LambdaRail across the United States [2]. The main goal of the project is to achieve large-scale deployment of diversified testbeds that support a federation of heterogeneous networks. It is part of the clean-slate movement aimed at exploring new design architectures for the future Internet. Because GENI is too large to be studied in detail in this book, only an overview of its general concept is provided in this section.

3.1.1 Basic Description of GENI

In its initial phase, GENI is divided into multiple control clusters. In [2], a detailed overview of its different control clusters as well as its slice control and management (C&M) mechanism is provided. The basic requirements of GENI are based on two main concepts: *resource sharing through virtualization* and a *federation ecosystem* of different testbeds [4]. Experimentation and network prototyping are the main applications of the GENI testbeds. Thus, the virtualization architecture of GENI is mostly focused on the *slicing* and *isolation* of different experimentations. The concept of federation is required to support concurrent research and collaboration among participating research projects. Different testbeds are designed with different applications and objectives in mind. One of the goals of GENI is to experiment with new *meta-management* frameworks to federate the different testbeds together.

GENI is also designed to support a set of advanced functionalities, including *deep programmability* (through SDN and virtualization), *advanced instrumentation system* and *advanced security features*. In fact, GENI is described as a *suite of research infrastructure* in the spiral two overview document [4]. The project is currently following the spiral model with each spiral lasting one year representing an engineering cycle from design prototyping to integration. The first spiral is aimed at fast prototyping of a large number of different projects and architectures. The second spiral is aimed at the sustainable integration of different architectures for long term experimentation [4]. The third spiral's goal is to improve usability and enrich existing tools and services. The spiral four is aimed at moving away from prototyping in order to increase support for real experimentations [5]. Currently, a large amount of simultaneous research projects are active and federated into the framework. As a federation of different testbeds, each of its control clusters is in itself based on a project of considerable scale, such as PlanetLab (Sect. 3.2) and ORBIT (Sect. 5.6).

3.1.2   GENI Conceptual Design

As explained in [4], the generalized GENI testbed meta-architecture can be divided into three main sections: the federation of aggregates of virtualized components, the clearinghouse management registries and the experimentation tool services [2, 4]. The component aggregates are the main resource pools in which multiple components are grouped under the control of an aggregate manager. The physical resources are located in these pools and are shared through virtualization managed by the local aggregate managers. Different aggregate managers are federated under the same generalized control framework [4]. The clearinghouse registries provide the book-keeping and security features required for user authentication and slice configuration. The different aggregates must communicate and establish trust relationship with the clearinghouse before allocating resources to a user. Finally, the tools and services contain the important software developed for the GENI users to monitor, control and debug their experiments. Using the terminology explained in [2], a *slice* represents a particular experiment and is formed by "stitching" together multiple *slivers* across the GENI framework. A *sliver* is defined as a container of a particular resource obtained either through virtualization or partitioning [6]. A diagram illustrating the interaction between these sections is assembled in Fig. 3.1.

The control framework shown in Fig. 3.1 is only a high-level abstraction. The actual implementation of the framework may vary. According to [4], the different GENI projects are grouped under multiple control clusters, each implementing

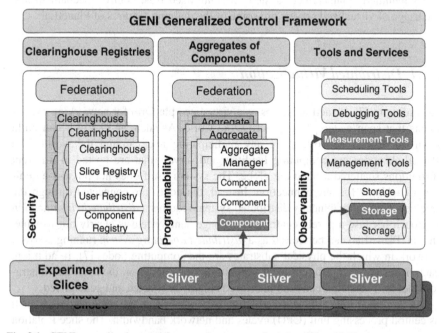

Fig. 3.1   GENI generalized control framework prototype (based on [2] and [4])

a version of the generalized meta-management framework with their own unique features. These clusters will not be discussed in greater details in this chapter, with the exception of PlanetLab in the following section. The ORBIT cluster includes implementations of wireless virtualization and will be briefly mentioned in Chap. 5. In terms of fitting wireless virtualization into the GENI terminology, a virtualized wireless testbed can be considered a part of an aggregate. Thus, wireless resources can be managed by an aggregate manager. This is the equivalent of a local wireless resource manager that can handle the sharing of the wireless resources and communicate with its corresponding global registries. Some of these requirements will be revisited in the Chap. 6. Similar requirements are also present in SAVI, as will be discussed in Sect. 3.4.

3.2 PlanetLab

PlanetLab, initiated in 2002, is one of the earliest successful large-scale *distributed* virtualized testbeds. It has served as an important source of experience for projects such as GENI [2]. As a result, some of its virtualization concepts can be found in the GENI control framework as one of its control clusters. Even if wireless virtualization is not part of PlanetLab, its *distributed virtualization* framework remains an interesting concept applicable to wireless virtualization. The PlanetLab report [7] and [8] provide an overview of general design principles concerning slice creation, slice management and resource allocation and scheduling. According to [8], an "unbundled" management system, a centralized trust control mechanism and the usage of virtualized servers are some of the defining features of PlanetLab.

3.2.1 Distributed Virtualization

According to [7], PlanetLab serves as a virtualized platform for networking experiments that span *geographically* across the world. As such, an end user can build a virtual testbed network using any participating node around the world. The PlanetLab nodes are computing facilities and servers that are hosted in different institutions scattered around the globe. The system itself does not possess a dedicated backbone for interconnection. Instead, the nodes are connected through the Internet. This geographically distributed system requires a *distributed virtualization* approach because a purely centralized management system is clearly not feasible. In fact, PlanetLab can be viewed as a *distributed operating system* running across a network in which the physical substrates are the computing nodes [7]. In addition, each nodes can run different local virtual machines based on the local administrator's needs. Thus, the PlanetLab interface is highly abstracted and generalized.

As described in [7], the slice isolation is performed on the node resources such as central processing unit (CPU) cycles and network bandwidth. The slice isolation mechanism must also isolate namespace and system privileges of competing slices on the same node. Furthermore, since the backbone is the Internet, safety mechanisms to limit resource consumption and prevent run-away effects of certain experiments on the local production network are set in place. The "unbundled" management system described in [7] and [8] consists of allowing *independent local management systems* to be integrated into the testbed through a common generic interface provided by the PlanetLab OS. The operation of each individual node is completely independent from each other. On the other hand, the *trust system* is *centralized* in the PlanetLab Central (PLC). The PLC acts as an intermediary broker between the users and the independent local node managers (NM). According to [2] and [8], the PLC supports two methods of slice instantiation. *The direct method* involves the user contacting the PLC which then directly communicates with the node manager and slice creation services located on each requested nodes on behalf of the clients. The *indirect* or *delegated method*, also known as Emulab, allows a service outside of the PLC to request *ticket tokens* from the PLC [2]. Then, the delegate services can instantiate new slices by exchanging the ticket with the node manager of requested nodes. The PlanetLab architecture will be described in the following subsection.

3.2.2 PlanetLab Architecture

A PlanetLab node is based on server virtualization using Linux VServers [9]. Each node, represented by a host computer, is running a Linux-based virtual machine monitor (VMM), or *hypervisor*, which supports different VServer VMs [8]. The node manager, which runs with root privileges on top of the VMM, is the main agent that allocates resources to new slices. Since independent management privileges must be given to each node, all operations must be performed through *services* running on the VMM [7]. Slices allocated on the node can either be *privileged*, *partially privileged* or *unprivileged*. The node manager is privileged as root whereas services that can call the node manager are partially privileged. The end users are unprivileged. The concept is to only grant the *minimum* set of privileges required for the functioning of each service. The core functions of PlanetLab are handled through the node manager whereas additional policies and configurations are handled inside the local administrator accounts. The overall architecture is shown in Fig. 3.2.

The resource allocations of both bandwidth and CPU cycles are handled using Linux kernel modules. Isolation between network traffic is realized by manipulating the Linux safe raw sockets [8]. Some existing issues with PlanetLab are mentioned in [2]. For instance, PlanetLab cannot efficiently run routing protocol experiments as it requires kernel-level privileges that are not directly available on a PlanetLab node [2]. Another main issue is the lack of repeatability of the experiment due to limited control over the Internet backbone. Nevertheless, multiple extension projects exist to compensate for these shortcomings. For instance,

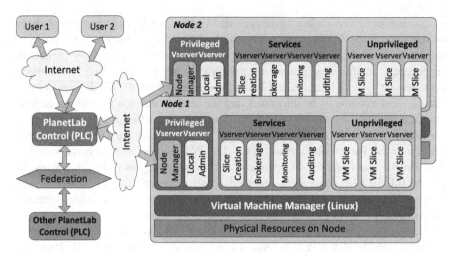

Fig. 3.2 PlanetLab system architecture (based on [7] and [8])

Emulab allows a more controlled environment by enabling portions of the experiment to be realized through emulation [2]. Finally, as shown in Fig. 3.2, different PLC instances can be federated to enable *customized development* of PlanetLab for specific applications. For instance, OneLab from FIRE is aimed at integrating wireless sensors into PlanetLab Europe [2]. Overall, the PlanetLab OS is heavily leveraging from functionalities implemented as Linux kernel modules.

In terms of wireless virtualization, it is interesting to note that a localized manager, such as the node manager in PlanetLab, can be used as the management interface for the virtualization of wireless access devices. The unbundled management system is excellent for a network of heterogeneous wireless technologies with different applications. The usage of node managers and local administration is a possible solution to handle heterogeneous resources by allowing service differentiation. The virtualization agents in VANI nodes, which will be presented in Sect. 3.4, apply similar strategies. These management concepts will be revisited in the context of wireless virtualization in Chap. 6.

3.3   European Initiatives in Network Virtualization

The Future Internet Research and Experimentation (FIRE) Project from Europe has similar specifications as GENI. Thus, detailed discussion of FIRE projects are not provided in this chapter. Just like GENI, FIRE is based on the federation of different experimentation networks across Europe [2], including PANLAB, OneLab, FEDERICA and numerous other initiatives. Many of their projects were part of the Seventh Framework Programme (FP7) of the European Union and are sharing technologies with GENI. As an example, OFELIA is a FP7-funded

OpenFlow-based testbed using the Expedient web interface control software from GENI [10]. GENI Expedient is an open-source software implementation of the GENI control interface written in Django, a Python-based web interface framework, and running on Apache2 servers [11]. It can be integrated with multiple GENI aggregates (as well as non-GENI testbeds) which can be supported through *extensible plugins*. The registries and slice information are stored in a structured query language (SQL) database. OFELIA is grouped into five "islands" that use Expedient to manage their OpenFlow-enabled network resources. More information on OpenFlow will be found in Sect. 4.1.

3.4 Smart Application on Virtual Infrastructure

Smart application on virtual infrastructure (SAVI) is an experimental testbed under development by the joint effort of Canadian industry and academia. In some sense, SAVI can be considered the Canadian counterpart of GENI. SAVI puts a high emphasis on *cloud infrastructure,* flexible high-speed *wireless access networks* and *low energy footprint.* The goal of SAVI is to address the design of future application platform built on a flexible infrastructure consisting of heterogeneous resources. Such an infrastructure will evolve and repurpose itself. It will also be able to support experimentation on the deployment, maintenance and retirement of future Internet applications [3]. SAVI is initially based on an extension of the virtualized application networking infrastructure (VANI), a virtualization testbed implemented by University of Toronto. The main goal of this section is to expose SAVI's design principles and its meta-management architecture.

3.4.1 Virtualized Application Networking Infrastructure

VANI is introduced by researchers from University of Toronto in [12] as a *service-oriented* virtualization platform for networking experiment. As opposed to PlanetLab which only offers processing nodes as resources, VANI is designed to allow various types of *heterogeneous resources* to be shared. Largely inspired by PlanetLab and GENI, VANI nodes aim to provide an experimentation platform for future networking development [12]. The VANI architecture served as a foundation for developing SAVI.

The VANI nodes can be deployed over dedicated raw Ethernet backbone, IP-based Internet backbone or a mixture of both. The nodes can also be connected through the Canadian optical fiber research network CANARIE [13]. This research-dedicated network infrastructure will allow VANI nodes to perform experiments without the same disturbances suffered by PlanetLab. In order to support non-IP experiment, special headers implementing *Q-in-Q tagging* can be used across an Ethernet substrate network [12]. On the other hand, the connection across the Internet is realized through gateways. The gateways must perform appropriate network address translation (NAT) to map the internal resource addresses into external IP addresses accessible to the user who requested these resources [12]. Each VANI node also has a set of VLAN tags to isolate the different slices of resources located on the same node.

According to [12], the VANI architecture is divided into two main planes: the *control management plane* and the *application plane*. The control management plane (CMP) is implemented using Business Processes Execution Language (BPEL), a web services-oriented XML-based language. Thus, the interface to the CMP is accessible through web services, similar to Expedient from GENI. The CMP performs the tasks of book-keeping and high-level resource management. Any user requesting a new slice of resources must contact the CMP. The types of resources supported on VANI nodes are *processing resources, programmable resources* (FPGA), *storage resources* and *fabric resources* [13]. Each resource type communicates with the CMP through a common generic interface written in Web Service Description Language (WSDL) [12]. A virtualization layer is customized for each type of resources and independent *virtualization agents* are located on the physical resources. The application plane is simply the set of user application running on top of the virtualized resources. The architecture is summarized in Fig. 3.3.

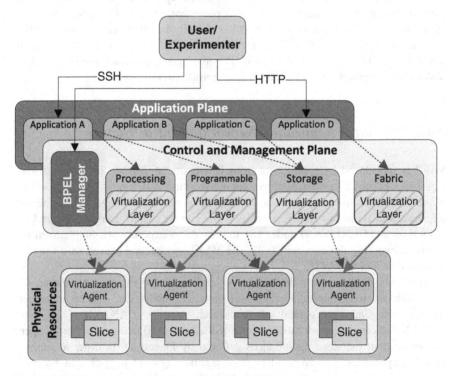

Fig. 3.3 VANI architecture model (based on [12] and [13])

Different types of resources are virtualized differently and independently from each other. According to [12], the *processing resources* are virtualized using instances of Linux VServers, similar to PlanetLab. The VMs can be connected to end users through secure shell (SSH). The *programmable resources* are BEE2 FPGA boards with Linux distribution running on embedded PowerPC general purpose processor (GPP) on which the *local virtualization agent* (LVA) is run. In order to allow testbed users to directly access the programmable resources, a hardware description language (HDL) interface module must be included in the users' HDL project [13]. The *storage resources* are volume servers running Hypertext Transfer Protocol (HTTP) services to allow direct connection between the user and the servers once the trust relationship is established through the CMP. The *fabric resources* include the switches, gateways and bridges. In the initial version of the VANI nodes, the switches are configured through the application layer-based simple network management protocol (SNMP) [13]. In the SAVI upgrade to VANI, the virtualization of the network fabric is realized through the integration of OpenFlow (Sect. 4.1). Overall, even though VANI nodes do not include wireless resources, it provides a *common interface* and *framework* for a various types of resources to be virtualized through customized approaches. The modular design of the web services and the virtualization agents presented in VANI are possible interfaces that can help the integration of wireless virtualization into a network infrastructure virtualization framework.

3.4.2 SAVI Architecture Model

The SAVI testbed was initially based on extending VANI. According to the SAVI internal report, the major components of SAVI are the *network fabric*, the *control and management center* and the *resource nodes*. The resource nodes are divided into three levels: core, edge and access. Each core node is composed of a large-scale datacenter possibly located close to a source of renewable energy such as a hydroelectric power station. A core node has large-capacity storage and cloud computing resources. On the other hand, edge nodes are smaller local datacenters hosted in participating universities. They can contain non-computing resources (such as programmable resources) in addition to a relatively scaled-down version of the storage and processing resources. Access nodes are similar to edge nodes with the exception that they are more focused on virtualized wireless and optical access resources. The integration of wireless access technologies is considered a major requirement of SAVI.

An important design decision of the SAVI framework inherited from VANI is the separation of the testbed into two planes: the control plane and the application (or data) plane. Both planes are software layers overlaid on the *physical resource plane*, which represents the hardware backbone of the virtualized testbed, as shown in Fig. 3.3. However, SAVI is more sophisticated than VANI by

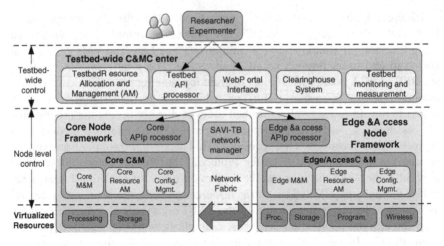

Fig. 3.4 Simplified SAVI testbed architecture

integrating a more elaborate control framework very similar to that of GENI. The
simplified conceptual diagram of the SAVI testbed, illustrating the different management entities, is shown in Fig. 3.4. The *testbed-wide C&M center* is a suite of
software that includes a clearinghouse system, a testbed monitoring and measurement (M&M) system, the resource allocation system and the web portal server.
The control center can be integrated within the core node but can also exist as a
separate entity. Each core and edge node also has its own local C&M system. The
management framework is based on the OpenStack® open source cloud computing platform [14]. A network management system is used to virtualize the network
fabric through OpenFlow and FlowVisor, which will be covered in Sect. 4.1 of this
book. The development of SAVI is being carried out in five distinct themes. Each
theme focuses on a particular aspect of the SAVI testbed such as smart applications, cloud computing, smart edge, integrated wireless/optical access and testbed
integration.

3.5 Project AKARI

For the AKARI Project from Japan, the document [15] presents the main specifications of different virtualization architectures from an abstracted perspective. Participating testbeds include the Japan Gigabit Network (JGN-X) from
the National Institute of Information and Communications Technology (NICT).
One virtualization project running over JGN-X is the Virtual Node Project
(VNP). Just like its counterparts GENI, FIRE and SAVI, AKARI relies on optical grid as backbone. Thus, it can take advantage of Japan's advanced optical
fiber network.

3.5.1 Alternative Virtualization Meta-Architectures

The initial work on AKARI examined a few interesting virtualization concepts and identified some important issues related to virtualization. According to [15], the AKARI project shares the same view as GENI in the sense that it favors a *competition model* in which different competing architectures are deployed simultaneously using the same virtualized resources. Traditional virtualization concepts are based on *isolated virtualization*, in which virtual networks sharing the same hardware are isolated from each other. Alternative virtualization architectures proposed by [15] include *transitive/cooperative virtualization* and *layered virtualization*. Transitive and cooperative virtualization implies a meta-architecture that can orchestrate several different virtual networks. The end user, instead of being assigned directly to a single virtual network, is now handled by the meta-architecture, which manages multiple virtual networks. This architecture is *fault tolerant* because the failure of one virtual network can be mitigated by offloading users to another virtual network. The transitive meta-architecture supports the *sustainable evolution* of different network architectures, allowing smooth transitions between them. The third meta-architecture, layered virtualization, allows the management of different independently-virtualized layers. The different meta-architectures are shown in Fig. 3.5. The document [2] suggests that a *clean-slate* approach should consider these different architectures.

From Fig. 3.5, it can be noted that the layered meta-architecture is similar to the separation of the testbed-wide control plane and the node-level control plane of SAVI in Sect. 3.4. On the other hand, the transitive meta-architecture is akin to a more advanced version of testbed federation.

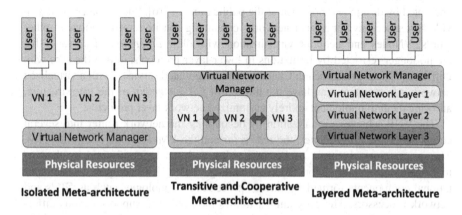

Fig. 3.5 Comparison between alternative virtualization meta-architectures (based on [2] and [15])

3.5.2 General Considerations for Virtualization

Some of the basic issues related to virtualization mentioned in [15] include the *implementation* of the virtualization layer, the *autonomy level* of subsystems and the *scaling* of the network. These issues are present in most virtualization architectures, including those for wireless virtualization. The implementation of the virtualization layer is a difficult process that will require insight in the particular technology that is being virtualized. In this process, the architecture of the control framework and the layer on which the virtualization hypervisor reside must be determined. The difficulty of optimization in resource management can vary depending on the granularity of control and the level of autonomy required [15]. Finally, the scaling of the meta-architecture can be limited by the latency and bandwidth bottlenecks of the system. Some of these issues will become a recurrent theme in later chapters in the context of wireless virtualization.

3.5.3 Virtual Nodes Network Virtualization Project

One of the goals of AKARI is to design an *evolvable* testbed that can quickly adapt to emerging networking research projects. One of the platforms implemented on the JGN-X testbed is the Virtual Node (VNode) Project (VNP) [16]. Each VNode is a virtualized router instance or slice on the JGN-X infrastructure. The implementation is based on a combination of existing protocols such as generic routing encapsulation (GRE) tunnels, virtual local area network (VLAN) and multiprotocol label switching (MPLS). As such, it can support regular virtual private network (VPN) functionalities compatible with existing non-virtualized routers. However, as opposed to traditional VPN, VNodes can allow the virtual network to run entirely new network protocols alongside existing ones. Possible applications supported on VNodes but impossible to implement on existing networks would be a proprietary non-IP protocol that is encrypted by design and not open to the public domain [16]. Such protocol will provide better security for virtual networks containing sensitive information. In other words, VNodes and network virtualization will allow *differentiated services* that are no longer restricted by limitations in lower layer protocols, unlike the case with DiffServ [17]. As mentioned in Chap. 1, the differentiated services, with a virtualized network infrastructure, can have the illusion of *full ownership* of the infrastructure, allowing them to implement entirely customized network layer protocols.

The VNP achieves virtualization by decoupling the network router into several components: the *programmer*, the *redirector* and the *VNode manager* (VNM) [18]. The programmer performs packet processing and can be configured by the virtual operators of the network slice. According to [19], it is composed of computational components formed by various processing technologies such as GPPs and network processors. The programmer allows the testbed to support various different protocols. The redirector provides an *independent* and *hidden* mapping of the

virtual network topology. This additional level of indirection and isolation has aim
to increase the modularity of testbed, removing any dependence to a particular
implementation of the programmer at the expense of incurring more overhead [18].
The redirector also performs the conversion of the data packets between the *internal*
VNode protocols and the *external* experimental slice protocols. Finally, the VNode
manager controls the policies and configurations of the VNode for the infrastructure
owner or *domain controller* (DC). It plays a similar role as the node manager from
PlanetLab (Sect. 3.2) or the local C&M in SAVI (Sect. 3.4). Overall, the VNode
implementation shares similar concepts with OpenFlow, which will be presented
in Sect. 4.1. It enables SDN by redirecting flows to an external software controller. However, unlike OpenFlow, VNP provides an extra layer of indirection which
allows the programmer to execute arbitrary functions not restricted to only routing
and switching. In fact, OpenFlow can be implemented *within* a VNode slice [18].

References

1. J. Pan, S. Paul, R. Jain, A survey of the research on future internet architectures. IEEE
   Commun. Mag. **49**(7), 26–36 (2011)
2. S. Paul, J. Pan, R. Jain, Architectures for the future networks and the next generation
   Internet: a survey. Comput. Commun. **34**(1), 2–14 (2010)
3. A. Leon-Garcia, NSERC strategic network on smart applications on virtual infrastructure.
   University of Toronto, Sep 2011. Available: http://www.savinetwork.ca/wp-content/uploads/
   Al-Leon-Garcia-SAVI-Introduction.pdf
4. Group-GENI, GENI spiral 2 overview. GENI-INF-PRO-S2-OV-1.1, Jun 2010. Available:
   http://groups.geni.net/geni/attachment/wiki/SpiralTwo/GENIS2Ovrvw060310.pdf
5. Group-GENI, GENI spiral four. Accessed on 26 Apr 2013. Available: http://groups.geni.net/geni/
   wiki/SpiralFour
6. Group-GENI, GENI glossary. Jan 2013. Available: http://groups.geni.net/geni/wiki/GeniGlossary
7. A. Bavier, M. Bowman, B. Chun, D. Culler, S. Karlin, S. Muir, L. Peterson, T. Roscoe, T. Spalink,
   M. Wawrzoniak, Operating system support for planetary-scale network services, in *Proceedings
   of the 1st Symposium on Networked Systems Design and Implementation*, March 2004
8. L. Peterson, A. Bavier, M.E. Fiuczynski, S. Muir, Experiences building PlanetLab, in *Proceedings
   of the 7th USENIX Symposium on Operating Systems Design and Implementation*, Nov 2006
9. Linux VServer Project, Overview. 22 Apr 2013. Available: http://linux-vserver.org/Overview
10. Group-OFELIA, Deliverable 5.1—first version of the OFELIA management software.
    OFELIA-ICT-285365, Mar 2011. Available: http://www.fp7-ofelia.eu/assets/Public-Deliverables/
    D5-1-final.pdf
11. J. Naous, P. Kazemian, R. Sherwood, S. Seetharaman, N. McKeown, G. Parulkar, G.
    Appenzeller, Expedient: A centralized pluggable clearinghouse to manage GENI experi-
    ments, Technical report, Jan 2010. Available: http://www.openflow.org/downloads/technicalre
    ports/openflow-tr-2010-1-expedient.pdf
12. H. Bannazadeh, A. Leon-Garcia, Virtualized application networking infrastructure. TridentCom
    2010, May 2010
13. K. Redmond, H. Bannazadeh, P. Chow, A. Leon-Garcia, Development of a virtualized appli-
    cation networking infrastructure node, in *Proceedings of the IEEE Workshop on Enabling the
    Future Service-Oriented Internet*, Dec 2009
14. OpenStack Cloud Software, Open source software for building private and public clouds.
    Available: http://www.openstack.org/

15. Network Virtualization, New generation network architecture AKARI conceptual design (ver2.0). Aug. 2009. Available: http://akari-project.nict.go.jp/eng/concept-design/AKARI_fulltext_e_preliminary_ver2.pdf
16. A. Nakao, Virtual node project—virtualization technology for building new-generation networks. NICT News issue no. 393, Jun 2010, pp. 1–6
17. S. Blake, D. Black, M. Carlson, E. Davies, Z. Wang, W. Weiss, An architecture for differentiated services, IETF RFC 2475, Dec 1998
18. Y. Kanada, K. Shiraishi, A. Nakao, High-performance network accommodation and intra-slice switching using a type of virtualization node. INFOCOM 2012, Oct 2012
19. Y. Kanada, K. Shiraishi, A. Nakao, Network-virtualization nodes that support mutually independent development and evolution of node components, in *Proceedings of 13th IEEE International Conference on Communication Systems (ICCS 2012)*, Oct 2012

Chapter 4
Network Virtualization Technologies and Techniques

This chapter continues with the discussion of network virtualization implementations and enabling technologies applicable within the context of wireless virtualization. More specifically, recent developments in software-defined networking (SDN) technologies are presented. Overall, SDN concepts are used to relocate functionalities from hardware into software by providing configurable interfaces. This increases the programmability of the hardware and facilitates the implementation of virtualization. First, OpenFlow, a SDN protocol for the network fabric that can enable *flow-based virtualization*, is covered along with its associated technologies. The application of OpenFlow in wireless access points is also briefly discussed. Then, in the second part of this chapter, SDN technologies for wireless networks, such as the control and provisioning of wireless access points (CAPWAP) protocol and CloudMAC, are examined. Finally, in the last part of this chapter, network interface card input/output (I/O) virtualization used in computer virtualization is presented. This is motivated by the fact that similar I/O virtualization techniques can be applied to wireless access devices.

4.1 OpenFlow and Software-Defined Networking

One of the main goals of SDN is to enhance the *programmability* of the network fabric and allow the *decoupling* of functionalities from hardware. OpenFlow is a SDN-enabling technology that allows multiple networking experimentations to be conducted simultaneously on the existing production network without disturbing its regular operation [1]. In short summary, OpenFlow is a protocol that allows an external controller to dynamically modify the forwarding table of switches and routers. OpenFlow is part of the Programmable Open Mobile Internet 2020 (POMI 2020) project initiated by Stanford University. It has since then gained its footing in project such as GENI and OFELIA, as well as support from the industry through the Open Networking Foundation (ONF). For instance, Google has revealed that the internal datacenter connection of their G-scale backbone network is successfully running OpenFlow [2].

H. Wen et al., *Wireless Virtualization*, SpringerBriefs in Computer Science,
DOI: 10.1007/978-3-319-01291-9_4, © The Author(s) 2013

As proposed in [1], OpenFlow can achieve network-level *abstraction* by providing an interface protocol to manipulate and isolate the traffic flows of switches and routers. The white paper [1] proposes OpenFlow as a protocol to centralize the control and setup of data flows in a network, effectively enabling multiple virtualized networks to coexist on shared physical resources. By itself, OpenFlow only exposes the networking functionalities of network nodes as software interfaces, in line with the spirit of SDN. However, these software interfaces can then be used by an external centralized controller to implement virtualization. Thus, network virtualization can be viewed as one of the potential applications of OpenFlow SDN. An OpenFlow hypervisor, such as FlowVisor presented in [3] and [4], can be used to virtualize the network. To run a particular network experiment, an OpenFlow controller such as NOX [4] can be deployed over the hypervisor as a *network operating system*. A typical network virtualization architecture using OpenFlow is shown in Fig. 4.1. The different components are explained in each of the following subsections. It is important to note that OpenFlow is a rapidly evolving technology standard. The OpenFlow concepts presented in this chapter are mainly based on the version 1.0 of the standard.

4.1.1 Basic Overview of the OpenFlow Protocol

OpenFlow is initially designed to manipulate the forwarding tables in switches and routers [1] using software *external* to the hardware firmware. The standard protocol defines a common set of low-level operations for switches and

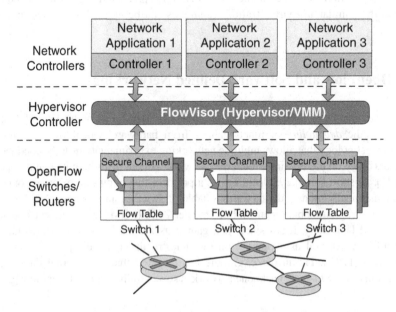

Fig. 4.1 Network virtualization model using OpenFlow (based on [1], [3] and [4])

router. As described in [1], an OpenFlow-compliant switch must have three components: a flow table, a secure channel and support for the OpenFlow protocol messages. In the version 1.0 of the standards defined in [5], the entries of the flow table contain three information fields: the header field, the flow counter field and the flow action field. A particular "flow" is defined by any combination of the header subfields from layer 1 to layer 4 along with possible wildcard fields. The flow counters are divided into different granularities as there are 32-bit and 64-bit counters for each table, each flow, each port or each queue [5]. The three basic actions supported are: *forward to port*, *forward to controller* and *drop packet* [1].

The basic operation of the protocol is extremely simple. An ingress packet is first matched with the field headers of existing flow entries in the flow table. If a longest prefix match is found, the appropriate counter is incremented and a list of actions prescribed for that entry is executed. If the actions cannot be executed in order, an error message is triggered [5]. If a match is not found, the packet is forwarded to the controller through the secure channel using the OpenFlow protocol. The controller can then decide to create a new entry in the flow table with a set of prescribed actions. In summary, the semi-persistency of rules setup in the flow table by the controller allows the packets to be processed at line rate while being managed by a *centralized* controller. The protocol effectively decouples the *data plane* from the *control plane*. The version 1.0 of the standard only supports basic features with IPv4 and Ethernet packets. Additional features are added in each new version of the standard as extensions. For instance, version 1.1 adds support for multicast and multiple flow tables per switch [6]. More recent versions have added support for IPv6 and other packet formats. Currently, the version 1.0 remains the most widely supported version with very basic features.

4.1.2 Basic Overview of OpenFlow Controllers

An OpenFlow controller software platform usually offers a set of application programming interfaces (APIs) for developers to use the OpenFlow commands to exert SDN. A single controller can be connected to a network of switches, each having its own datapath identification (DPID). One of the first OpenFlow controllers is NOX [3]. In some sense, NOX can be viewed as a network operating system using OpenFlow as its instruction set [3]. From the SDN developer's point of view, NOX is an interface platform on which customized *controller applications* can be run, similar to a computer operating system. In other words, NOX provides *network abstraction* [4]. It offers a set of APIs and event call-backs to *facilitate* the programming of an OpenFlow controller. The main logical components of NOX consist of *events*, *namespace* and *network view* [4]. These elements are the main building blocks for controller applications. Basic OpenFlow events, such as *packet-in*, are treated as interrupts that can trigger call-back handlers in controller applications. The original NOX, called NOX-Classic, is implemented in a mix of Python and C++. Recent development of NOX has split the controller into a pure-C++ implementation called NOX and a pure-Python implementation called POX [7]. NOX/POX is one of the numerous open-source controller software platforms currently available. Other open-source controllers exist, such as the Java-based Beacon [8] and Project Floodlight [9], and the Python-based Ryu [10]. Different controllers have varying degrees of supported features depending on the targeted application. For instance, Ryu has integrated support for the OpenStack® cloud computing virtualization software platform [11], facilitating its deployment in testbeds such as SAVI (refer to Sect. 3.4).

4.1.3 Basic Overview of OpenFlow Virtualization

Network virtualization with OpenFlow can be achieved by using an OpenFlow hypervisor such as FlowVisor. FlowVisor is a Java-based controller that enables *multiple* OpenFlow controllers to share the same physical resources. It is a useful mechanism to allow multiple virtual network operators to control and manage their own slice. Its isolation features include bandwidth, topology, traffic, computational resources and forwarding tables [3]. Each slice allocated by the hypervisor is associated to a particular *flow-space* and a *guest controller* (GC). A flow-space is defined by a set of flows characterized by any combination of the header fields. Each guest controller is allocated a separate ID and separate queuing buffers. The hypervisor keeps track of the flows associated with each slice in a similar fashion as a clearinghouse registry. In order to use FlowVisor, neither the guest controller nor the network switches need to be modified (other than support standard OpenFlow) [3]. In other words, the FlowVisor acts as a transparent *intermediate* between the guest controllers and the network elements, as shown in Fig. 4.1. From the perspective of a guest controller, it has full ownership over the network slice it has been allocated. From the perspective of the network nodes, the FlowVisor acts as the unique controller. In such a way, the guest controllers are effectively abstracted from the network elements and vice versa. The FlowVisor also performs conflict resolution among slices. For instance, a flow table modification signal sent by a guest controller is intercepted by the FlowVisor and modified accordingly to prevent conflicts with another guest controller. Some of the limitations of the FlowVisor mentioned in [3] include cross-layer overhead during flow setups. The FlowVisor APIs are implemented using the JavaScript Object Notation (JSON). Resource allocation algorithms are developed separately from the FlowVisor as extension modules. However, only a common version of OpenFlow across all guest controllers is supported in the current FlowVisor. Different guest controllers with different OpenFlow versions are not supported.

To remove some of these limitations, alternative OpenFlow virtualization methods are described in [12] and shown in Fig. 4.2. In general, the hypervisor is viewed as a *translation* or *mapping* unit between the virtual network and the physical hardware. The main design decisions for OpenFlow virtualization are: the location of the hypervisor, the depth of resource partitioning, the number

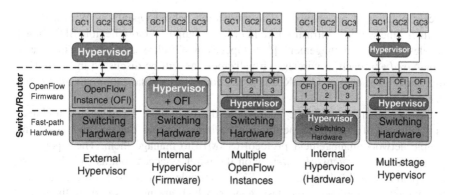

Fig. 4.2 Alternative network virtualization models using OpenFlow (based on [12])

of OpenFlow client instances and the type of configuration channel [12]. The OpenFlow hypervisor can be implemented either as an external controller, inside the firmware located on the switching hardware or in both (*multi-stage* or *multi-layer* hypervisor). Corresponding to where the hypervisor is located, resource partitioning and slicing can be performed at the controller level, at the OpenFlow firmware level or integrated within the low-level switching fabric. At the controller level, the flow-space can be logically partitioned by the hypervisor by intercepting and overwriting OpenFlow commands of guest controllers, just as in FlowVisor. However, since flow-spaces cannot overlap (a "flat" space definition), FlowVisor exhibits a lack of flexibility [12]. On the other hand, at the firmware level, the flow tables can be partitioned *hierarchically*. After the introduction of multiple flow tables in OpenFlow 1.1 [6], a *chain* of linked flow tables suggested in [12] can be used as a branching structure for virtualization. A master ingress table allows flows to branch out to different table chains using a go-to command while a master egress table performs the final multiplexing [12]. Each chain of tables can then be managed by a different guest controllers. Alternatively, there can be a single OpenFlow plugin inside the switch firmware or multiple instances of OpenFlow clients using separate connections with corresponding guest controller. Finally, the control and configuration channel can be in-band or out-of-band. In the case in-band channel is applied, appropriate quality-of-service (QoS) mechanisms must be implemented to guarantee prioritized access for configuration messages over data traffic [12]. Enhancements to the basic OpenFlow protocol to support advanced QoS functionalities are suggested in [12].

4.1.4 OpenFlow Wireless (OpenRoads)

The success of OpenFlow in SDN applications encouraged various extensions to the original standard to support new functionalities. OpenFlow Wireless, also known as OpenRoads, is the addition of OpenFlow-enabled plug-ins on wireless

access points and basestations. The motivation behind OpenFlow Wireless is to
"unlock" the wireless infrastructure in order to support a more flexible SDN-based
wireless network management [13]. OpenFlow Wireless can be classified as a
flow-based approach to wireless virtualization, as will be discussed in Chap. 6 of
this book. The setup of OpenRoads in the context of virtualization is summarized
in Fig. 4.3.

As shown in Fig. 4.3, the addition to the OpenFlow virtualization architecture consists of enabling OpenFlow on wireless access devices and using the
simple network management protocol (SNMP) to manage them. The OpenFlow
plug-in allows the *network gateway* on the wireless access points to behave like
an OpenFlow-enabled switch. The SNMP module is used to handle configuration parameters unique to wireless devices, which are not covered by the standard OpenFlow protocol. For instance, OpenRoads can use SNMP to configure the
power level, frequency, data rate and service set identifier (SSID) of 802.11 access
points [13]. The SNMP Visor isolates wireless device configuration instances
from different slices of the network. FlowVisor and SNMP Visor work in conjunction as the hypervisor layer of OpenRoads. Potential applications include mobile
handover between Wi-Fi and WiMAX demonstrated in [13]. The implementation
presented in [13] contains multiple *alix3d2*-based 802.11 access points (APs) and
a WiMAX basestation, both with modified firmware support for OpenFlow. The
802.11 APs use an embedded GPP running a Linux-based firmware distribution
running Open vSwitch (OVS), an OpenFlow-enabled software switch [14]. OVS
is typically used in datacenter servers to provide network virtualization between
VMs. In this case, it acts as a SDN-enabled network gateway on APs, such that
flows can be regulated *before* entering the wired network.

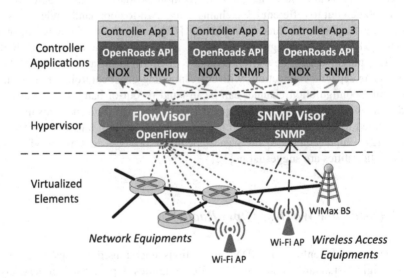

Fig. 4.3 OpenFlow wireless architecture with virtualization (based on [13])

Unfortunately, as stated in [13], there are limitations with OpenRoads. First, OpenFlow cannot affect any wireless functionality, leaving all configuration tasks to the SNMP. However, SNMP is a generic management protocol not sufficiently specialized to handle all wireless functionalities. The level of shareable wireless configuration is limited to what can be implemented by the developer. For instance, conflicting configurations such as different power levels on the same wireless access point cannot be resolved. Thus, efficient wireless resource allocation and interference management cannot be achieved without a more organized management framework or more enhancements in the radio hardware. Such enhancements will be discussed in Chaps. 5 and 6 of this book. As an alternative, management protocols specific to wireless devices such as control and provisioning of wireless access points (CAPWAP) or RFC 5415 [15] can replace SNMP (refer to Sect. 4.2). Nevertheless, OpenRoads provides a good start to bridge wireless virtualization with SDN and network virtualization.

4.1.5   OpenFlow and Other SDN and Virtualization Technologies

OpenFlow can be considered as a form of datapath virtualization technology, as discussed in Chap. 2. Other widely-known datapath virtualization technologies include virtual local area network (VLAN) and multiprotocol label switching (MPLS). VLAN is traditionally used to transform a single physical switch to multiple virtual switches by partitioning the interface ports and tagging the packets. On the other hand, MPLS is designed to provide flexible routing by appending protocol-independent labels on packet flows. It has since then evolved into a network virtualization technique used in link and network layer-based virtual private networks (VPN) [16]. Unlike these datapath virtualization techniques, basic OpenFlow operations do not rely on the *labeling* of flows, allowing them to travel through the network fabric *unmodified*. Thus, OpenFlow is more flexible and extensive than most other datapath virtualization technologies. In fact, OpenFlow can be applied on VLAN, VPN and MPLS networks. For instance, version 1.1 supports the pop, push and swap functions of MPLS [6]. Additionally, OpenFlow is compatible with other virtualization frameworks and technologies, such as OVS [14] and the OpenStack software [11]. The OpenStack technology is an open source cloud management platform initially formed by three main software components: Nova, Swift and Glance [17]. Nova manages VMs for cloud computing with VM images provided through Glance whereas Swift manages virtualized storage. A fourth component dedicated to networking, Neutron [18], is added in order to support extensible and differentiated network applications and services over a virtualized cloud infrastructure. Support for network virtualization through OpenFlow is integrated in Quantum and implemented in testbeds such as SAVI (refer to Sect. 3.4).

Of course, OpenFlow is not necessarily the magical solution to all the problems that arise in SDN and network virtualization. In fact, it has its own challenges, such as scalability and fault-tolerance. A purely centralized model might suffer from scalability and bottleneck issues because of the overhead and delay during each flow setup. Thus, special controllers and extensions have been made. For example, DevoFlow [19] improves the performance and scalability of the controller by removing unnecessary level of network visibility and reducing the number of flow entries and flow control messages. On the other hand, in HyperFlow [20], *distributed* controllers exchange *local network state* information among them through a publish/subscribe process to maintain a *global network state*. Implemented through NOX, HyperFlow decouples the frequent local network events from the infrequent global network events. However, because of the propagation of state information distributed among HyperFlow controller nodes, one important challenge is to keep the convergence time of a consistent global network state stable and bounded [20].

Other important technologies that can assist SDN and network virtualization include the network configuration protocol (NETCONF). Standardized in RFC 4741 and 6241 [21], NETCONF is based on remote procedure calls (RPC) and extensible markup language (XML). It provides a more intuitive and flexible configuration interface than SNMP. Whereas OpenFlow controls the data forwarding plane, NETCONF allows the dynamic configuration of network equipments. In other words, OpenFlow and NETCONF can complement each other in the SDN deployment. For instance, the management of OpenFlow switches and routers can be performed through NETCONF. In the case of wireless networks, there is CAPWAP, which is discussed in the following Sect. 4.2.

4.2 Software-Defined Networking in Wireless Networks

Whereas OpenFlow and other datapath virtualization technologies enable advanced SDN and flow-based network virtualization, they lack capabilities to control functionalities unique to wireless technologies, such as spectrum planning in a wireless network. This section covers some SDN and network management technologies specifically applied to *wireless networks*.

4.2.1 Control and Provisioning of Wireless Access Points

The control and provisioning of wireless access points (CAPWAP) protocol is a control and management (C&M) protocol defined in general in IETF RFC 5415 [15] and for IEEE 802.11 in RFC 5416 [22]. It is based on the lightweight wireless access point protocol (LWAPP) or RFC 5412 [23]. CAPWAP is aimed at relocating certain wireless functionalities from the local hardware equipment to an external centralized controller. It *centralizes* the C&M of AP functionalities in order to support a software-defined architecture. Thus, CAPWAP share similar concepts with SDN technologies such as OpenFlow. It can be used to implement *network functions virtualization* [24] for wireless hardware. The original CAPWAP protocol is technology-independent, potentially allowing the same control framework to be deployed over wireless nodes with *heterogeneous technologies*. The implementation of CAPWAP on a specific wireless technology is called a *binding* [25]. The 802.11 WLAN binding is discussed in this subsection. Bindings to other wireless technologies are not defined as of the writing of this book, although a similar architecture can be applied.

CAPWAP splits the access point into two components: the *access controller* (AC) and the *wireless termination points* (WTPs) [26]. The role of the AC is to implement customized C&M functions for wireless networks, which is similar to the role of the OpenFlow controller in Ethernet networks. The WTP is a local agent installed on the access point hardware that communicates with the AC. As opposed to OpenFlow, CAPWAP offers wireless device configuration capabilities such as channel selection and multiple SSIDs. It is also able to manage the 802.11e QoS for Wi-Fi multimedia (WMM) by modifying the QoS parameters of 802.11 MAC frames. There are two different operational modes: *Local MAC* and *Split MAC* [22]. In Local MAC, only high-level management and configuration functions, such as client authentication and security settings, are processed at the AC. On the other hand, in Split MAC, most non-time-critical functionalities, including encryption and QoS classification, can be relocated into the AC [22]. The two control architectures are illustrated in Fig. 4.4.

CAPWAP protocol defines two main types of messages: control messages and data messages. Control messages include WTP discovery messages, configuration request messages and configuration setup messages [25]. Data messages can

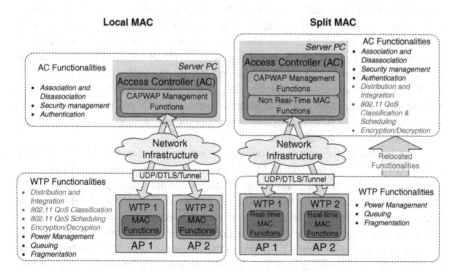

Fig. 4.4 Operational modes for IEEE 802.11 binding of CAPWAP (based on [15], [22] and [26])

include the payload (in Local MAC) or the entire 802.11 MAC frame (in Split MAC) encapsulated and tunneled to the AC for *remote processing*. The control path and the datapath are established over separate UDP ports using Datagram Transport Layer Security (DTLS) [15]. The interaction between the AC and the WTP can be represented by a finite state machine (FSM). The detailed operation of the protocol is not covered here and can be found in [15]. An open-source version of CAPWAP, OpenCAPWAP, is presented in [25]. OpenCAPWAP supports both operational modes and is compatible with Linux-based AP firmware such as OpenWrt [27]. As alternatives to CAPWAP, various proprietary implementations of *network controllers* have also been developed to manage enterprise-class WLANs. Thus, the concept of centralized wireless network management is clearly not new.

Once again, CAPWAP by itself is not a virtualization technology, similar to the situation with OpenFlow. However, CAPWAP does provide a mechanism to decouple the wireless functionalities from the physical hardware to a *remote* and *centralized* software controller. The next step required for virtualization would be the design of a CAPWAP hypervisor in which multiple virtual network operators can manage their own slice of the network using CAPWAP, just as in the case of guest controllers in FlowVisor.

4.2.2 Wireless Network Virtualization with CloudMAC and OpenFlow

According to [28], some of the motivations behind the shifting of functionalities from the local wireless firmware to a centralized controller are to reduce management complexity and accelerate innovation. In the traditional wireless network architecture, the addition of new features and functionalities often makes the wireless firmware overly bloated. To address these issues, protocols such as CAPWAP were developed to keep the frontend *lightweight*. However, CAPWAP has a few shortcomings. Notably, the synchronization between the finite states of the AC and the WTPs can be extremely complex [28]. Additionally, CAPWAP by itself does not support virtualization. To address these issues, CloudMAC is proposed in [28] as an alternative WLAN architecture that supports not only SDN but also a form of wireless network virtualization.

The CloudMAC is an OpenFlow-based architecture for 802.11 WLAN that allows APs to forward the 802.11 MAC frames to cloud-based servers for processing. It is composed of four main components: the virtual access points (VAPs), the wireless termination points (WTPs), the OpenFlow-enabled network fabric and an OpenFlow controller. The architecture is summarized in Fig. 4.5. The VAPs are VM instances located in the cloud that contains one or multiple *virtual* wireless network interface cards (NICs). Similar to the AC in the Split MAC architecture of CAPWAP, the VAP handles all wireless functions typically processed in a wireless firmware, such as authentication, association and beacon frames generation [28].

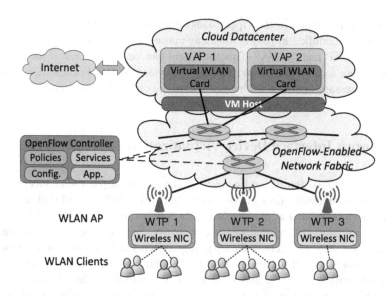

Fig. 4.5 CloudMAC network architecture (based on [28])

Thus, a client station is logically associated with the VAP and not the WTP, giving rise to a more flexible mobility management. The WTPs are based on stripped-down lightweight AP firmware. They handle low-level *time-critical* functions such as the distributed coordination function (DCF), acknowledgement and retransmission. It receives and forwards all frames to the VAP for processing. A single VAP can be connected to multiple WTPs as the centralized controller of the WLAN. However, the interesting feature is that a single WTP can also be connected to multiple VAPs, differentiated by their (virtual) MAC address. This effectively enables the WTPs to be shared among multiple VAPs, facilitating network virtualization [28]. The WTPs and VAPs are connected through layer 2 tunnels running over the OpenFlow-enabled network fabric [28]. No state synchronization is required between the VAP and the WTP, unlike in CAPWAP. In addition, no modifications to the client station are required.

In CloudMAC, OpenFlow is a key technology used to increase the *flexibility* of the WLAN. With OpenFlow, the binding between the WTPs and VAPs can be *dynamically* reconfigured, allowing *on-demand* allocation of new WTPs to a VAP. The OpenFlow controller is used as an infrastructure-wide hypervisor. It manages control and management frames and stores network slicing policies. It reinforces these policies by intercepting and overwriting control headers exchanged between the WTPs and the VAPs. Custom infrastructure-wide network applications and services can be implemented on the controller. Some possible applications suggested in [28] include dynamic spectrum usage and on-demand APs. In [28], the WTPs are implemented using OpenWrt while the VAPs are Debian VMs hosted on the VMWare vSphere® virtualization platform. The tunneling protocol is Capsulator

[29], which was applied in the OpenRoads setup in [13]. One of the limitations of CloudMAC identified in [28] is the delay caused by the round-trip time (RTT) between the WTP and the VAP. This limitation can be problematic if the cloud-based VAPs are located too far away from the physical WTPs. For instance, client association messages must be exchanged within the order of milliseconds.

4.3   Network Interface Device Virtualization

In a virtualized cloud infrastructure, a server can host multiple VMs connected to different virtual networks through virtualized network interface card (NIC). Since NICs are devices located at the interface between the host computer and the network infrastructure, NIC virtualization can involve both computer virtualization techniques (for device drivers in the operating system) and network virtualization techniques (for low-level firmware on the NIC). Although the focus of this book is not on host virtualization, it is important to note that wireless access devices, especially for 802.11 WLAN, operate using *wireless* NICs, which share many common points with Ethernet NICs. Thus, there is a striking similarity between wireless access nodes and host NICs in the sense that they are both *endpoint access devices* to the network infrastructure. This section provides a simple overview of NIC virtualization architectures: single-root input/output virtualization (SR-IOV) and Crossbow. In contrast, the virtualization of wireless network NIC will be discussed in Sect. 5.6 of Chap. 5.

4.3.1   Basic Overview of Single-Root I/O Virtualization

Input/output (I/O) virtualization is an important feature used by the virtual machine manager (VMM) or hypervisor to share I/O interfaces located on a host machine among VM instances. SR-IOV is a hardware-based virtualization technology for Peripheral Component Interconnect Express (PCIe) interfaces. According to [30], the intervention of the hypervisor in data transfers is one major source of performance degradation for I/O virtualization. Thus, one feature of SR-IOV is to push the I/O memory and address translation from the hypervisor to the *I/O memory management unit* (IOMMU). In other words, instead of performing the translation in hypervisor software, the translation is performed directly through the IOMMU, a *hardware-based* component. The other interesting characteristic of SR-IOV is its separation of *virtual functions* (VF) from *physical functions* (PF). The physical functions represent the fully-featured *master* mode of PCIe functionalities. Each PF can support multiple VF instances with *lightweight* or *reduced* functionality. Each VF has its own set of *performance-critical* resources but must share the general device-wide resources, such as PHY processing and packet classification [30]. One virtualization architecture design that uses SR-IOV in

Fig. 4.6 NIC virtualization architecture using SR-IOV (based on [30])

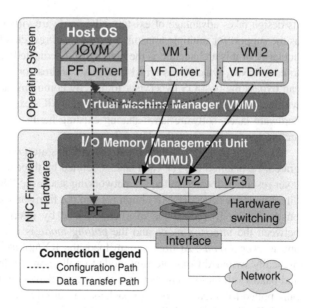

conjunction with the Xen® hypervisor and kernel-based virtual machine (KVM) is proposed in [30] and is illustrated in Fig. 4.6.

As shown in Fig. 4.6, the VF drivers can *directly* communicate with the VF instances located on the hardware device, *without* the intervention of the VMM and the PF driver. The host OS is contacted only when a guest VM modifies the configuration of its interface. In that case, the *I/O virtualization manager* (IOVM) is used to control the PF instance located on the hardware through the PF driver. The PF instance is in charge of provisioning new VFs or modifying the configuration of existing VFs. In other words, the PF driver is only responsible for the configuration and management of VF instances during VM setup, which are expected to be infrequent compared to data transfers. Most of the virtualization of time-critical functions is performed in hardware, parallel to the less-demanding management functions performed in software. Thus, the virtualization is distributed across both software and hardware. In the particular architecture proposed by Intel in [30], additional optimization techniques involving interrupts were employed to enhance the performance of SR-IOV but they are not covered in this book. According to [30], the scalability and efficiency of their architecture is very high due to the optimizations, with only 2–3 % overhead for each additional VM and more than 50 % reduction in CPU usage compared to SR-IOV without these custom optimizations.

4.3.2 Virtualized Network Interface Cards Using Crossbow

Crossbow, proposed by the Oracle Solaris Kernel Networking Group in [31], is an OpenSolaris™ network stack virtualization architecture based on virtual NICs (VNIC).

Its design takes advantage of existing NIC hardware virtualization technologies, such as the support for multiple transmit and receive *rings* (circular buffers). The main particularity of this architecture is the binding of a dedicated hardware *virtualization lane* to each VNIC, which is assigned to a MAC address or VLAN tag. Then, the VNIC can bypass the host OS hypervisor to directly communicate with their dedicated lane with no contention [31]. When the number of VNICs exceeds the number of available hardware lanes, software lanes sharing existing hardware lanes are provisioned. The architecture is summarized in Fig. 4.7.

As shown in Fig. 4.7, the virtualization lanes are *cross-layer* and *vertically* integrated. As detailed in [31], each lane is defined by three sets of resources: the MAC-layer resources located in the OS networking stack, the NIC driver resources such as direct memory access (DMA) bindings and the firmware/hardware resources located on the NICs. To reduce the overhead caused by interrupts, a dynamic switching between the *interrupt mode* and the *polling mode* is implemented through *polling* and *worker threads*. At high system load, the RX/uplink networking stack is automatically switched from interrupt mode to polling mode. In order to achieve *fairness* and bandwidth partitioning between the RX virtualization lanes, each receive ring is only polled for a fixed amount of packets per period consistent with the assigned link speed. For the TX/downlink traffic, direct flow control mechanism is implemented as network filters in the system. One potential disadvantage of Crossbow is its reliance on multiple cores to process polling and worker threads simultaneously in order to reduce processing overhead. According to [31], there is no measurable performance degradation between a physical NIC and a virtual NIC using a dedicated hardware lane. Since wireless access devices are similar to NICs, similar cross-layer virtualization strategies can be used for wireless virtualization. Such approaches will be discussed in the context of software-defined radio (SDR) in Sect. 5.7 of Chap. 5.

Fig. 4.7 Crossbow architecture and virtualization lanes (based on [31])

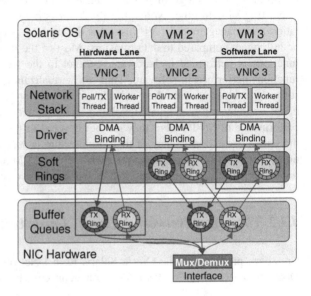

References

1. N. McKeown, T. Anderson, H. Balakrishnan, G. Parulkar, L. Perterson, J. Rexford, S. Shenker, J. Turner, OpenFlow: Enabling innovation in campus networks. ACM SIGCOMM Comput. Commun. Rev. **38**(2), 69–74 (2008)
2. R. Merritt, Google describes its OpenFlow network. EE Times News and Analysis, 17 Apr 2012. Available: http://www.eetimes.com/electronics-news/4371179/Google-describes-its-OpenFlow-network
3. R. Sherwood, G. Gibb, K.-K. Yap, FlowVisor: A network virtualization layer. OpenFlow Switch Consortium, Technical Report, Oct 2009. Available: http://www.openflow.org/downloads/technicalreports/openflow-tr-2009-1-flowvisor.pdf
4. N. Gude, T. Koponen, J. Pettit, B. Pfaff, M. Casado, N. McKeown, S. Shenker, NOX: Towards an operating system for networks. Comput. Commun. Rev. **38**(3), 105–110 (2008)
5. OpenFlow switch specification version 1.0.0. Dec 2009. Available: http://www.openflow.org/documents/openflow-spec-v1.0.0.pdf
6. OpenFlow switch specification version 1.1.0, Feb 2011. Available: http://www.openflow.org/documents/openflow-spec-v1.1.0.pdf
7. Nox. Available: http://www.noxrepo.org/
8. Beacon home. Available: https://openflow.stanford.edu/display/Beacon/Home
9. Project floodlight open source software for building software-defined networks. Available: http://www.projectfloodlight.org/floodlight/
10. Ryu. .Available: http://osrg.github.io/ryu/
11. OpenStack Cloud Software, Open source software for building private and public clouds. Available: http://www.openstack.org/
12. P. Sköldström, K. Yedavalli, Network virtualization and resource allocation in OpenFlow-based wide area networks, in *Proceedings of 2012 IEEE International Conference on Communications*, Jun 2012
13. K.K. Yap, R. Sherwood, M. Kobayashi, T.-Y. Huang, M. Chan, N. Handigol, N. McKeown, G. Parulkar, Blueprint for introducing innovation into wireless mobile networks, in *Proceedings of the Second ACM SIGCOMM Workshop on Virtualized Infrastructure Systems And Architectures*, Sep 2010
14. Open vSwitch—an open virtual switch. Available: http://openvswitch.org/
15. P. Calhoun, M. Montemurro, D. Stanley (eds.), Control and provisioning of wireless access points (CAPWAP) protocol specifications. IETF RFC 5415, Mar 2009. Available: http://tools.ietf.org/html/rfc5415
16. A. Wang, M. Iyer, R. Dutta, G. Rouskas, I. Baldine, Network virtualization: Technologies, perspectives, and frontiers. J. Lightwave Technol. **31**(4), 523–537 (2012)
17. OpenStack, Quantum and open vSwitch—part I. Available: http://openvswitch.org/openstack/2011/07/25/openstack-quantum-and-open-vswitch-part-1/
18. OpenStack Neutron. Available: https://wiki.openstack.org/wiki/Neutron
19. A.R. Curtis, J.C. Mogul, J. Tourrilhes, P. Yalagandula, P. Sharma, S. Banerjee, DevoFlow: scaling flow management for high performance networks. SIGCOMM-Comput. Commun. Rev. **41**(4), 254 (2011)
20. A. Tootoonchian, Y. Ganjali, HyperFlow: A distributed control plane for OpenFlow, in *Proceedings of the 2010 Internet Network Management Conference on Research on Enterprise Networking*, Apr 2010
21. R. Enns, M. Bjorklund, J. Schoenwaelder, A. Bierman (eds.), Network configuration protocol (NETCONF). RFC 6241, ISSN 2070-1721, Jun 2011. Available: http://tools.ietf.org/html/rfc6241
22. P. Calhoun, M. Montemurro, D. Stanley (eds.), Control and provisioning of wireless access points (CAPWAP) protocol binding for IEEE 802.11. IETF RFC 5416, Mar 2009. Available: http://tools.ietf.org/html/rfc5416
23. P. Calhoun et al., Lightweight access point protocol. IETF RFC 5412, Feb 2010. Available: http://tools.ietf.org/html/rfc5412

24. Network functions virtualisation: An introduction, benefits, enablers, challenges and call for action. Introductory White Paper, SDN and OpenFlow World Congress, Oct 2012. Available: http://www.tid.es/es/Documents/NFV_White_PaperV2.pdf

25. M. Bernaschi, F. Cacace, G. Iannello, M. Vellucci, L. Vollero, OpenCAPWAP: An open source CAPWAP implementation for the management and configuration of WiFi hot-spots. Int. J. Comput. Telecommun. Networking **53**(2), 217–230 (2009)

26. L. Yang, P. Zerfos, E. Sadot, Architecture taxonomy for control and provisioning of wireless access points (CAPWAP). IETF RFC 4118, Jun 2005. Available: http://tools.ietf.org/html/rfc4118

27. OpenWrt wireless freedom. Available: https://openwrt.org/

28. P. Dely et al., Cloudmac—an Openflow based architecture for 802.11 MAC layer processing in the cloud, in *Proceedings of 2012 IEEE GLOBECOM Workshops*, Dec 2012

29. Capsulator. Available: http://www.openflow.org/wk/index.php/Capsulator

30. Y. Dong, X. Yang, X. Li, J. Li, K. Tian, H. Guan, High performance network virtualization with SR-IOV, in *Proceedings of the IEEE 16th International Symposium on High Performance Computer Architecture (HCPA)*, Jan 2010

31. S. Tripathi, N. Droux, T. Srinivasan, K. Belgaied, Crossbow: From hardware virtualized NICs to virtualized networks, *in Proceedings of the 1st ACM Workshop on Virtualized Infrastructure Systems and Architectures*, Aug 2009

Chapter 5
Wireless Virtualization

Wireless virtualization can be considered as an umbrella term for *wireless access* virtualization, *wireless infrastructure* virtualization, *wireless network* virtualization and even *mobile network* virtualization. In general, wireless virtualization can be interpreted as the *sharing* and *abstraction* of wireless access devices among multiple users or user groups with a certain degree of isolation between them. Similar to the case with the term network virtualization, the term wireless virtualization is very broadly defined. Thus, this chapter is focused on presenting the different *aspects* of wireless virtualization such that the readers can acquire a better understanding of the context in which it is applied. The different aspects discussed in this chapter include the role of multiple access techniques in virtualization, the integration of wireless resources in testbed architectures, the applications enabled by a virtualized wireless infrastructure and the virtualization of different wireless technologies. The link between wireless virtualization and software-defined radio (SDR) technologies is also highlighted. A more detailed analysis and definition of wireless virtualization will be provided in the following Chap. 6.

In this chapter, it is important to distinguish the different levels of *resource allocation* used under different contexts. In wireless virtualization, a resource can refer to the wireless equipment such as the entire basestation as well as low-level PHY resources such as space–time–frequency slots. Some of these resources can be *shared* whereas others can only be *partitioned*. The abstraction of low-level resources provides the illusion that high-level resources (such as entire basestations) can be shared although low-level resources must ultimately be partitioned to support that sharing. First, an overview of some basic aspects of wireless virtualization is made in Sect. 5.1. This is followed by a discussion of the use of multiple access and multiplexing techniques in wireless virtualization. Then, the subsequent sections cover the following topics: general wireless virtualization frameworks that can be integrated within a virtualized network infrastructure, user-centric and cloud-centric applications for wireless virtualization, implementations of wireless virtualization in cellular networks and wireless local area

H. Wen et al., *Wireless Virtualization*, SpringerBriefs in Computer Science, DOI: 10.1007/978-3-319-01291-9_5, © The Author(s) 2013

networks (WLANs), SDR technologies in virtualization and the application of radio spectrum slicing techniques. In short, general architectures, concepts and applications of wireless virtualization are presented in the first half of this chapter whereas wireless virtualization architectures in the context of specific wireless technologies are presented in the second half.

5.1 Different Aspects of Wireless Virtualization

In order to provide a more structured survey of the different wireless virtualization concepts, a basic overview of different aspects of wireless virtualization is presented in this section. Nevertheless, since wireless virtualization is a young and active area of research, a comprehensive survey of all the recent advances in wireless virtualization technologies is quite difficult to realize. Overall, wireless virtualization can be considered a *multi-dimensional* concept with many aspects to consider. This section focuses on three major aspects: the *depth* of virtualization, the virtualization of different *wireless technologies* and the virtualization of *infrastructure-side* or *client-side* technologies. Note that the majority of wireless virtualization architectures and implementations often involve more than one aspect of wireless virtualization.

5.1.1 Scope and Depth of Virtualization

The first aspects discussed in this section are the *scope* and *depth* of a given virtualization architecture or technology. In terms of *scope*, research projects on wireless virtualization can range from generic frameworks to implementations of specific virtualization techniques. On one hand, high-level perspectives focus on wireless network management, general wireless virtualization design guidelines and abstraction interfaces to integrate wireless resources into a virtualized network infrastructure (refer to Sect. 5.3). They apply a *network-wide scope*. On the other hand, low-level perspectives typically explore the virtualization of resources within each *individual* wireless node. In other words, they apply a more *localized scope*.

The *depth* of virtualization is the extent of penetration of slicing and partitioning on the wireless resources. It is tied to the *granularity* of virtualized resources and often dictates where the hypervisor is located inside the virtualization architecture. For example, in the *overlay flow-based virtualization* (refer to Sect. 6.1), network links and traffic flows are virtualized. In that case, the hypervisor can be a network filter sitting on top of the wireless networking stack. However, the depth of virtualization is very low and does not allow multiple protocol stacks to share the same hardware. With a *deeper* virtualization, the full wireless protocol stack can eventually be sliced, as in the case of SDR-based virtualization, which will be discussed in Sect. 5.7. In some sense, the scope and depth of virtualization are tied to the *wireless virtualization perspectives*, a categorization framework presented in Sect. 6.1 of Chap. 6.

5.1.2 Virtualization in Different Wireless Technologies

As opposed to network virtualization technologies, which are mainly Ethernet-based, wireless virtualization can be applied to a larger variety of wireless access technologies and standards. Ethernet-based network virtualization architectures rarely have to address lower physical layer protocols. In the case of wireless virtualization, due to the relative scarcity of wireless resources, the unpredictable nature of wireless transmission and the multi-user multi-accessed wireless medium, the penetration of virtualization into the MAC and PHY layer protocols might be necessary to optimize wireless resource allocation. In addition, existing standards are constantly evolving and new applications are rapidly emerging, leading to the formation of a *heterogeneous network*. A virtualization layer that is applicable to multiple wireless standards can be combined with configurable SDR technologies to solve many of these emerging issues, which include maintaining QoS, smooth handover and hardware sharing across multiple wireless technologies.

Currently, there are three main categories of wireless technologies and standards based on the range of their application. The first category includes very short range wireless devices such Bluetooth and ZigBee (IEEE 802.15.4), which are used for *personal area networking* (PAN) and *sensor networks*. Then, the second category includes short range devices dominated by the wireless local area network (WLAN) IEEE 802.11 family of standards, commercially certified as Wi-Fi. The third category includes the medium and long range devices, such as wireless metropolitan area network (WMAN) technologies like WiMAX and cellular network technologies like Long Term Evolution (LTE). It is important to notice that not all these technologies can equally benefit from wireless virtualization. The benefits of virtualization are more apparent when the supported data rate and the supported number of users are relatively high, leaving room for possible sharing of resources on the infrastructure. Thus, 802.11 WLAN technologies and cellular technologies, which satisfy the requirements of high data rate and high number of users, are the main focus of most active research in wireless virtualization.

Cellular technologies have a natural affinity for infrastructure-level virtualization architectures due to their advanced quality-of-service (QoS) support, multi-user multi-access scheduling and network management capabilities. On the other hand, current 802.11 Wi-Fi technologies have a less elaborate control and management framework due to the *plug-and-play* nature of its setup. In WLAN, QoS is mainly based on the hybrid coordination functions (HCF) defined in 802.11e [1] whereas management can be realized through CAPWAP (refer to Sect. 4.2). Unfortunately, this QoS framework is only priority-based. Thus, unlike cellular technologies, no scheduling is supported in the native MAC. Clearly, due to these differences, the virtualization approach to wireless devices is *technology-dependent*. Nevertheless, a certain convergence between WLANs and cellular networks can be observed. The upcoming 802.11ac WLAN standard will support multi-user MIMO (MU-MIMO) [2], becoming similar to a miniaturized cellular basestation. On the other hand, the concept of *femtocell* or *small cell* networks makes cellular basestations self-configurable and easier to deploy, similar to WLAN APs. To take into account the possible convergence of different wireless technologies in the future, some wireless virtualization architectures aim at supporting heterogeneous networks. The abstraction provided through virtualization can potentially enable different standards to coexist on the same virtualized infrastructure.

5.1.3 Infrastructure-Side and Client-Side Virtualization

Wireless virtualization can also be referred to as *wireless access virtualization* or the virtualization of the *access* and *distribution* of data through the wireless medium. The access network is composed of the coordination and distribution points, called *basestations* (BS) in cellular terminology or *access points* (AP) in 802.11, and the *mobile clients*, called *user equipments* (UEs) in cellular terminology or *client stations* (STAs) in 802.11. The integration of virtualization technologies on both side of the access network is possible and has different applications. From the infrastructure perspective, both the *downlink* and the *uplink* of client users can be virtualized. Additionally, the concept of *user groups* [3] can only be implemented and coordinated from the basestations or access points. Thus, this chapter is mainly focused on infrastructure-side virtualization which requires no modifications to the existing client devices. Nevertheless, some applications, such as lossless wireless handover, advanced mobility management and optimized uplink allocation, might benefit from some modifications to the client-side interfaces and hardware. Finally, in ad hoc networks, mesh networks and machine-to-machine (M2M) applications, there is no clear distinction between basestation and client. A given wireless node must often perform *both* basestation and client functionalities. The virtualization of such mesh networks is briefly discussed in Sect. 5.6.

5.2 Multiple Access Techniques and Wireless Virtualization

Before continuing with the discussion of wireless virtualization architectures, it is important to identify the low-level resources involved in wireless transmission. The lowest level of wireless resources is based on the space–time-frequency access to the wireless medium. The allocation and multiplexing of these resources are referred to as *multiple access and multiplexing techniques*. These techniques consist of partitioning the time, space or frequency dimensions of the channel and allocating them to different users or traffic flows. The basic techniques include space-division multiple access (SDMA), time-division multiple access (TDMA), frequency-division multiple access (FDMA) and code-division multiple access (CDMA). Hybrid time–frequency methods such as orthogonal frequency-division multiple access (OFDMA) is also widely used in modern wireless technologies. Overall, multiple access techniques have an objective very similar to that of virtualization, which is to *share* the physical resources among *multiple* parties. In the case of the standard (non-virtualized) multiple access, the different parties are *individual users* or *QoS classes* of traffic. In the case of the virtualized multiple access, the different parties are *virtual network slices* or *groups of users*. Since these two concepts are fundamentally tied together, all wireless virtualization architectures ultimately rely on some combination of multiple access techniques, whether by design or as a by-product of the implementation. This section provides a brief discussion of the relationship between multiple access techniques and virtualization.

5.2.1 Slicing, Sharing and Flexibility in Wireless Virtualization

A preliminary study report by GENI [4] provides some basic comparison of the application of multiple access techniques in virtualization. According to [4], there is a difference between *slicing* and *virtualization* in the context of the wireless medium. The slicing of resources refers to the process of assigning a particular resource to be part of a network slice. Although the allocated resource of a slice can be virtualized, slicing does not *necessarily* imply virtualization and sharing of resources. In other words, a full radio node can be given to a slice without virtualization of the node itself. In that case, the process is called *resource partitioning*. On the other hand, the virtualization of a resource implies that multiple users or slices can *share* the *same* physical resource. Ultimately, all high-level resource virtualization eventually breaks down into low-level resource partitioning, as it is fundamentally impossible to truly share an irreducible physical resource. Of course, there is also a distinction between different degrees of sharing, in line with the discussion of the *scope* and *depth* of virtualization in Sect. 5.1. For instance, current cellular networks already support the concept of *network sharing* [5] in the sense that multiple network operators can share the same infrastructure. However, the underlying infrastructure has to follow the same wireless standard with a limited degree of customization. According to [6], a fully-virtualized and open infrastructure not only allows the sharing of the infrastructure but also allows different mobile virtual network operators (MVNOs) to deploy different protocol stacks over the same radio resources. One way to reach such level of flexibility is to use basic radio resource blocks as the *fundamental unit* of virtualization independent from the higher-layer architecture. Multiple access, multiplexing and spectrum slicing (refer to Sect. 5.8) techniques are used to allocate these fundamental resources.

5.2.2 Comparison of Multiple Access Techniques

Time, frequency, space and even code can all be used to partition the wireless channel resources. This subsection summarizes the different virtualization
approaches classified by [4], [6] and [7]. A generalized radio resources partitioning from [6] is represented in Fig. 5.1. As illustrated in Fig. 5.1, some degree of
orthogonality must be preserved both within and between the different dimensions. However, orthogonality is not a necessary requirement in practice. For
instance, practical *asynchronous* CDMA do not rely on orthogonal codes by
design. Spatial and frequency orthogonality can be difficult to achieve sometimes
due to signal interference and time orthogonality can be hindered by delays in the
signal (or the so-called *time interference*).

Each multiple access technique has its own set of advantages and disadvantages. They are used to *conserve* and *trade-off* different types of resources [7].
For example, pure FDMA can be used to assign each virtual network (VN) a
particular frequency, conserving the spatial resources by allowing multiple VNs
to reuse the same node [7] while consuming frequency resources. Unfortunately,
in the case of single carrier transmission, FDMA needs to be time-switched and
suffers from channel switching delay, which can be in the order of milliseconds
[4]. FDMA can also be limited by a crowded spectrum band with scarce availability of frequency resources. Alternatively, pure TDMA can be used to allocate each VN to a time slot but with the same frequency, conserving frequency
resources [7]. Depending on whether the integration of TDMA is pipelined,
TDMA virtualization can potentially suffer from additional context-switching
time in the order of milliseconds [4]. SDMA is a third possible technique presented in [4] that can refer to two different methods. One of them is to spatially
separate the allocation of full wireless nodes in an indoor wireless grid such as

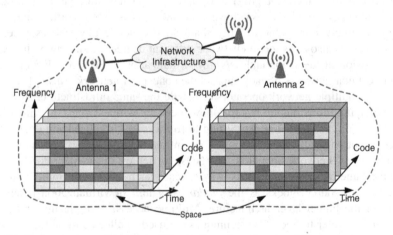

Fig. 5.1 Basic resource partitioning in different dimensions (based on [6])

ORBIT [7]. The other one is to allocate different spatial streams of a MIMO system to each VN. In the ORBIT testbed, the former one is applied to allow different experiments to run on different portions of the grid. One issue raised by [4] is the signal interference between nodes that are not sufficiently spatially separated. Another issue is the lack of flexibility since VNs must be spatially segregated, which is not achievable in practical scenarios outside of the testbed. Optionally, CDMA, a direct sequence spread spectrum access technique, can also be applied to support virtualization. While CDMA consumes more frequency resources (spread spectrum) it does not incur channel switching delays as in FDMA and TDMA. However, CDMA technologies are considered as legacy support in the new generation of cellular standards such as LTE.

Hybrid methods that use more than a single dimension can sometimes provide a more efficient and flexible multiplexing of VNs in virtualization. For instance, OFDMA is a multi-carrier multiple access technique that allows multiple users to seamlessly share different time–frequency blocks without any switching delay or interference penalties originally described in [4] and [7]. In OFDMA, inter-channel interference can be avoided by preserving orthogonality between resource blocks in frequency and inter-symbol interference can be minimized through the use of guard intervals and cyclic prefix between resource blocks in time. The downlink for new-generation LTE and WiMAX both uses OFDMA. The OFDMA scheduling is performed by the air interface MAC scheduler, which can be modified to integrate virtualization concepts, as will be discussed in Sect. 5.5. Overall, [4] provides a basic guideline on the usage of *different* multiple access techniques for *different* wireless applications depending on the requirements of the application. Thus, generic wireless virtualization frameworks *independent* of the particular type of radio resources being virtualized are presented in the following section.

5.3 General Architecture for Wireless Virtualization

This section presents two generalized control and management frameworks for the integration of wireless virtualization into a virtualized network infrastructure respectively proposed in [6] and [8]. Although these frameworks do not provide detailed implementations, they offer a global perspective on the design, deployment, control and management of the virtualization of the wireless network.

5.3.1 Wireless Virtualization as Mobile Network Virtualization

The framework proposed in [8] discusses the concept of *mobile network virtualization*. The concepts presented in this framework are mainly derived from network virtualization. In fact, the mobile network virtualization framework is viewed as an *extension* of network virtualization. The framework provides two main *design guidelines* which can be applied to a generic wireless virtualization framework: the partitioning of the architecture into functional *modules* or *building blocks* and the different levels of integration of the control plane.

First, [8] claims that any virtualization architecture can be broken down into three general entities: the virtualized physical resources (VPR), the virtual resource manager (VRM) and the virtual network controller (VNC). This classification fits the network virtualization meta-architectures presented in Chap. 3. The types of VPR considered in [8] include wireless access points and basestations. Each type of VPR has its own standardized interface since they can offer different types of services and functionalities. The VRM is the equivalent of the hypervisor layer or the virtual machine monitor (VMM) in computer virtualization. The VRM reinforces the QoS requirements of each slice determined by a network service-level agreement (SLA) negotiated between the slice owner and the infrastructure provider. It can also act as a *broker* of *aggregated resources*. Finally, the VNCs are used by the MVNOs to control and manage their own slice of the infrastructure. Overall, this architecture closely resembles other network virtualization frameworks, simply viewing wireless resources as an addition of a new type of resources to the *resource pool*. The overall architecture is shown in Fig. 5.2.

One interesting contribution of the [8] is the discussion about different *control schemes* applicable in a virtualized wireless infrastructure. The concept of *horizontal control* and *vertical control* is defined in the context of wireless virtualization. Horizontal control refers to the management across a group of common components and resources (i.e. among wireless nodes). On the other hand, vertical or *cross-layer* control refers to the management of resources across different network layers and domains (i.e. coupling between the cloud-based content, the network fabric and the wireless access). These control schemes are particularly relevant in a wireless infrastructure because QoS can only be guaranteed based on the interactions between the wired transport network and the wireless access network. Three examples of control applications that are enabled by a virtualized

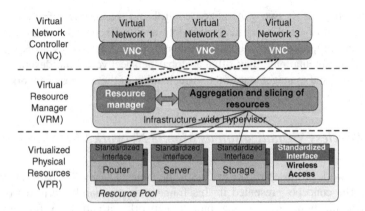

Fig. 5.2 Generalization of network virtualization to wireless virtualization (based on [8])

infrastructure are discussed in [8]. First, *end-to-end control* enables service delivery from the cloud-based content provider to the end user while keeping the entire transport and distribution process managed by the same framework. This allows the content and application provider to *extend* their applications and services into the infrastructure, if needed. The second application, *multilayer control*, consists of the cross-layer control of the data transport and distribution inside the infrastructure in order to reduce overhead. In other words, instead of segregating the control scheme of each layer, new cross-layer and cross-domain control frameworks can be deployed and collocated with the existing frameworks. Finally, the control of *heterogeneous network*, which consists of the abstraction and handover between different technologies, can be supported as a key feature for the ubiquitous access of the future wireless infrastructure. Overall, these control schemes provide some guidelines on how wireless access resources can be integrated and managed with the rest of the network infrastructure.

5.3.2 Configurable Radio Networks

In contrast with the framework presented in the previous subsection, which offers a global perspective of wireless resources within the network infrastructure, this subsection, based on [6], discusses a generalized virtualization framework specifically for wireless radio nodes. This framework emphasizes on the virtualization of low-level radio resources in order to provide full *configurability* of the wireless protocol stack within each slice. The architecture presented in [6] follows the same structural organization as [8] from the previous subsection with a few minor changes. First, the VPRs are uniquely composed of *radio resource blocks* delimited by time, frequency and code. In this case, the VRM of [8] is called the virtualization manager interface (VMI). The VMI is an autonomic and localized component on each wireless radio node that negotiates the allocation of the physical resource blocks with the MVNOs. It only acts as a coordinator. The actual interference and collision avoidance of radio resources between different slices is performed by the resource allocation control (RAC). The architecture is summarized in Fig. 5.3.

According to [6], the list of virtualization configuration parameters that should be supported in order to provide a flexible infrastructure includes both high-level network management functionalities and low-level radio PHY functionalities. Some of the high-level functionalities listed in [6] include routing, mobility management and congestion control, which can be jointly managed with the wired network. The physical layer functionalities consist of channel coding, multiplexing and antenna frontend management. As shown in Fig. 5.3, different virtual MAC and PHY layers can be constructed using the functionalities exposed through the framework. Finally, [6] also encourages the virtualization framework to be *autonomous* and *self-configurable*. It presents one possible mechanism for the automatic discovery of new radio nodes in the virtualized infrastructure. When an active

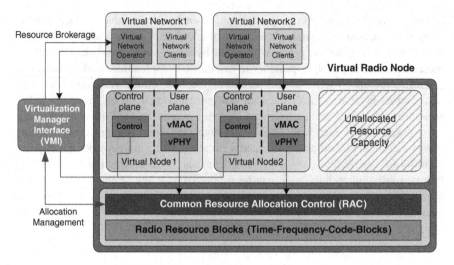

Fig. 5.3 General architecture for a single configurable virtual radio node (based on [6])

radio node is connected to the infrastructure, it advertises through the VMI information that allows the auditing of the node. The discovery of new nodes is then propagated to all MVNO clients.

Overall, [6] justifies the need for high configurability as necessary for the full *decoupling* of the physical infrastructure from the logical protocol, which is similar to how a general purpose processor can run arbitrary algorithms. On the other hand, communication networks are fundamentally specialized so it might be difficult to provide full abstraction of physical resources. While a full decoupling might be infeasible, advancement in such directions can lead to new business model for the deployment of new network infrastructure. One example given by [6] discusses the possibility of the virtualized wireless infrastructure being funded as part of the public infrastructure (public ownership) on which private MVNOs can operate with a cost. A more complete scenario of such futuristic infrastructure model will be provided in Sect. 6.2 of Chap. 6.

5.4 Example Applications of Wireless Virtualization

This section presents architectures that are applications dependent on the existence of a virtualized wireless network infrastructure. In other words, the following architectures can be implemented as services over a virtualized infrastructure. Their focus is on how virtualization can be exploited to support various applications in areas such as mobility management and ubiquitous wireless coverage. Thus, they are motivated by the assumption that the future of wireless technologies will be dominated by a *dense* and *ubiquitous* network. First, a cloud-centric mobility management architecture called Carmen is presented. Then, a virtualization-based user-centric solution to coordinated multipoint coverage in a cellular network is discussed.

5.4.1 Cloud-Centric Architecture for Rich Mobile Networking

The cloud-centric architecture proposed in [9] is aimed at the support of multi-dimensional mobility on a virtualized end-to-end service distribution network. In [9], the concept of a *mobile personal grid* (MPG) is explored. The MPG is defined as a *personal cloud* of mobile devices an user employs in his or her daily life. One of the goals of this architecture is to maintain a personal cloud that seamlessly accompanies the user. The emphasis is on the *multi-dimensionality* of the mobility. The user can switch between different devices such as smart phones and laptops. At the same time, the user can move across different contexts and environments (referred as *personas* in [9]) such as work and personal leisure. Different contexts and applications will imply different service requirements. Ultimately, the user can be mobile across different network infrastructures such as the enterprise WLAN network, the public cellular network and the local area network at home. In LTE terminology, some of these functionalities are managed through the mobility management entity (MME).

In order to support mobility across these dimensions, the Carmen framework is spread across three domains, as illustrated in Fig. 5.4: the user mobile devices, the virtualized wireless network infrastructure and the cloud service datacenters. The key component of the architecture is a cloud-based personal connectivity manager application called "Avatar" [9]. This personal assistant gathers resource usage statistics across all devices of the MPG and performs provisioning of handovers across different dimensions. Since this manager resides in the cloud, it has centralized visibility and access of user devices, which is necessary to maintain the MPG. As shown in Fig. 5.4, the architecture relies on the *virtualized* wireless network infrastructure to support *mobility* and *convergence* of access technologies. Interworking with SDN and virtualization technologies such as OpenFlow to slice the link resources has been proposed in [9]. Finally, the mobile devices owned by the end users must have a mobility client installed in order to manage connectivity, monitor usage patterns and report to the cloud-based Avatar.

The researchers in [9] have implemented a statistics monitoring service using the Carmen architecture. The prototype, named Clarinet, is a server-client application-based architecture spread across the mobile devices and the cloud, as shown in Fig. 5.4. The Clarinet cloud-based server hosts Avatar instances and a connectivity management database. It contains a web-based interface portal for personal access, management and configuration of accounts. The Clarinet client is installed on the mobile devices and sends usage statistics to the Clarinet server through the Extensible Messaging and Presence Protocol (XMPP), a XML-based messaging standard. In addition, the Clarinet client offers a set of APIs for other applications

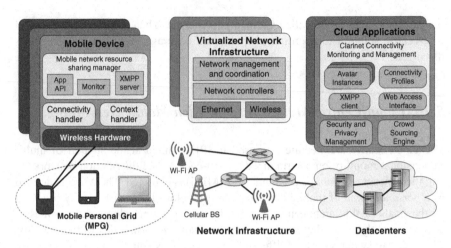

Fig. 5.4 Cloud-centric mobility management across different domains (based on [9])

running on the mobile device to access the mobility statistics and take advantage of the cloud-based Avatar service. Overall, the cloud-centric architecture is a proof-of-concept mobility-oriented application that leverages from the integration of cloud services into a virtualized wireless infrastructure [9].

5.4.2 User-Centric Architecture for Coordinated Multi-Point

Another possible application of a virtualized infrastructure, more specifically the cellular infrastructure, is the user-centric architecture for coordinated (or cooperative/competitive) multi-point (CoMP) proposed in [10]. The proposed architecture aims at leveraging advances in virtualization technologies to solve difficulties in the implementation of CoMP techniques. While [10] does not provide how wireless virtualization is implemented, it discusses some basic research directions for the design of such a framework.

The concept of CoMP transmission is to use multiple transmission points (TP) to improve the service quality and coverage of users in a cellular network, especially at the edge between two cells. It exploits diversity techniques in a cellular network that is increasingly growing denser. In a more traditional *cell-centric architecture*, multiple transmission points, or radio frontend, are handled by a baseband processing unit (BBU), as shown in Fig. 5.5. However, there are few technical difficulties with this approach. Notably, the PHY processing tasks, which include channel estimation, scheduling, clustering, beamforming, and decoding, must be jointly performed by two separate BBUs at the cell edge. Cell-centric CoMP attempts this synchronization through a master–slave protocol between the two BBUs of interest, leading to a complex and non-intuitive solution. The researchers in [10] argue that virtualization of the PHY functions of the physical BBU can create an alternative *user-centric architecture*. In this new scenario, illustrated on the right-hand side of Fig. 5.5, a more *virtual* user-centric coverage area is defined. Each user group is served by a *virtual BBU*, which can be spread across multiple physical BBUs. This decoupling of the physical BBU from the user groups obsoletes the complex master–slave interaction required in cell-centric CoMP. This paradigm shift also makes the joint algorithms more intuitive to implement through the abstracted virtual BBUs. In addition, virtual BBUs can be allocated on demand based on network load in order to support a more energy-efficient and flexible distribution network.

Of course, this architecture merely attempts to simplify the complex synchronization by decoupling processing functionalities from the physical hardware. In other words, it pushes the issue of synchronization between BBUs to the virtualization layer. According to [10], the feasibility of implementation of this architecture is supported by the presence of powerful multi-core general purpose processors (GPP) which are becoming increasingly common in wireless hardware equipment. Thus, many processing tasks performed by hardware digital signal processing (DSP) modules can be migrated to GPPs. This trend will accelerate the development of software-based programmable technologies for wireless hardware such as SDR. Nevertheless, there are technical issues that still need to be addressed. For instance, the design of the control channel for a virtual BBU can also prove to be challenging. The authors of [10] suggest a per-user independent virtual control channel in order to avoid interruption in the case of a handover. In

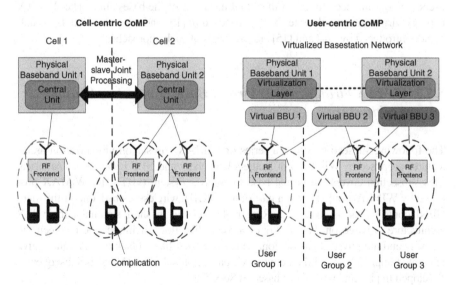

Fig. 5.5 Comparison between cell-centric and user-centric architecture for CoMP (based on [10])

terms of running a distributed algorithm across a virtualized infrastructure, there are scalability issues due to delays in the synchronization of the data. However, this bottleneck can be alleviated by restricting the exchange between BBUs to partially-processed lightweight metrics [10]. Finally, GPPs are not yet fully optimized for DSP operations and lead to larger overall power consumption [10]. Some of these issues are good examples of problems that will need to be solved in the design of a wireless virtualization framework for cellular networks.

5.5 Virtualization in Cellular Networks

As briefly mentioned in Sect. 5.1, the main difference between WLAN technologies and cellular technologies, other than their transmission range and data rate, is the presence or lack of elaborate *multi-user scheduling*. Unlike WLAN access points, cellular basestations are already subject to complex control and management functions. Thus, there exist a few different perspectives to take advantage of existing functionalities to virtualize a cellular network infrastructure. One of the approaches is to emulate the entire basestation in a VM instance, allowing each *virtual basestation* to manage the scheduling of its own users. In this case, the virtualization layer behaves like a scheduler of schedulers, handling the interaction between *groups of users* as opposed to individual users. The underlying hardware can be decoupled from the virtual basestations and even treated as a black box. The WiMAX virtual base transceiver station (vBTS) proposed in [11] and [12] falls into this category. Another approach is to modify the existing scheduling and QoS mechanisms to isolate and differentiate group of users, without the need to have full virtual instances of the basestation. The WiMAX network virtualization substrate (NVS) proposed in [13] and the LTE eNodeB virtualization introduced by [14] and [15] are examples of such approaches.

5.5.1 WiMAX Virtualization: Virtual Base Transceiver Station Architecture

The following subsections cover a series of WiMAX basestation virtualization architectures proposed by WINLAB of Rutgers University and NEC Laboratories America. This subsection covers the virtual base transceiver system (vBTS) for WiMAX proposed by WINLAB in [11]. It is a part of the ORBIT wireless testbed, one of the main GENI control clusters located at Rutgers University. In short summary, the vBTS architecture is based on the *overlay* of a router-like flow slicing engine between the gateway and the physical basestation, which is treated like a black box. An alternative integrated approach called the network virtualization substrate (NVS) is subsequently developed in [13] and will be discussed in Sect. 5.5.3.

The motivation behind vBTS is to enable multiple mobile virtual network operators (MVNOs) to share the same infrastructure [11]. Once again, the requirements of the system are common with those of GENI: programmability of the platform, observability of the measurements and repeatability of the experiments. According to [11], because the basestation hardware is proprietary, the virtualized basestation transceiver system (vBTS) can only exist as a framework located *outside* of the basestation. As a result, the vBTS architecture is a *portable software-based* solution located at the service gateway level. However, as a trade-off, the extent of slice isolation and radio access control is limited. At the same time, only downlink traffic can be virtualized due to the lack of control over the uplink transmission [13]. These issues are addressed by *integrating* the scheduler within the basestation in NVS (refer to Sect. 5.5.3).

A normal WiMAX basestation is composed of the base transceiver system (BTS), the access service network (ASN) gateway and the connectivity service network (CSN) gateway. Whereas the BTS controls transmission settings such as frequency, power and data rate, the ASN gateway connects the BTS to the wired network. The vBTS architecture consists of emulating a full basestation in VM instances and modifying the ASN gateway to *redirect* traffic coming from the newly-introduced *vBTS* to the physical BTS. The vBTS instances are implemented using full virtualization with Kernel-based virtual machine (KVM) in conjunction with Quick Emulator (QEMU), an user-mode hardware emulator. As shown in Fig. 5.6, each vBTS supports remote accessing through secured shell (SSH). All vBTS interface ports have VLAN tags. The traffic coming from each vBTS is intercepted by the slice isolation engine (SIE) located on the modified ASN

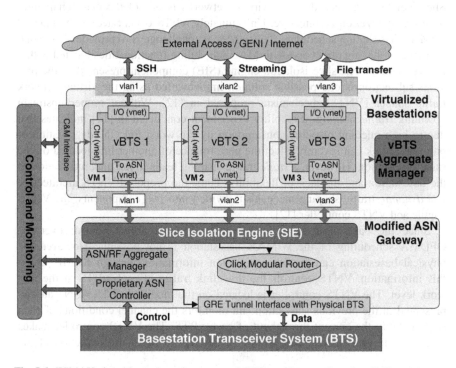

Fig. 5.6 WiMAX virtual base transceiver system (vBTS) architecture (based on [11])

gateway, acting as the virtualization layer. The SIE is implemented through the *virtual network traffic shaper* (VNTS) which will be covered in Sect. 5.5.2. It performs the task of closed loop-controlled traffic shaping to allocate bandwidth to each vBTS slice. The Click modular router software [16], a part of VNTS, then routes each traffic flow to the appropriate generic routing encapsulation (GRE) tunnel interface based on VLAN tags and MAC addresses. Finally, certain basestation configuration functionalities are accessed through a proprietary controller provided by equipment manufacturer. The overall control and management system is based on the GENI framework discussed in Sect. 3.1 of Chap. 3. One major advantage of this architecture is the flexibility of supporting different MAC schedulers in different vBTS since they are completely independent. However, the effectiveness of the scheduling modules is limited due to a lack of coupling with the real scheduler inside the physical basestation. As such, only a coarse isolation between different virtual slices can be achieved [13]. Since the virtualization is purely flow-based, the SIE can be enhanced by other flow-based virtualization alternatives, such as OpenFlow. The VNTS architecture for the SIE is introduced in [12] and discussed in greater details in the following subsection.

5.5.2 WiMAX Virtualization: Virtual Network Traffic Shaper

Slice isolation between different virtual networks is one of the main challenges of any virtualization architecture. Unfortunately, vBTS cannot access the internal WiMAX basestation schedulers. Thus, the main challenge is to provide slice isolation without having direct access to MAC scheduling functionalities located within the basestation. The slice isolation engine (SIE) component presented in the previous subsection is one possible solution implemented using the virtual network traffic shaper (VNTS) mechanism proposed in [12]. Without proper upstream downlink *traffic shaping*, a mobility experiment conducted in [12] indicates that due to high queuing resource consumption by the worst link, the degradation of one virtual link can severely affect the performance of another link located on the same basestation. However, a *static* traffic shaping leads to the underutilization of the basestation and the lack of flexibility. Thus, the proposed VNTS architecture is a *dynamic* traffic shaping mechanism divided into two components: the VNTS engine and VNTS controller [12].

Since the basestation is a black box component, direct control over the OFDMA scheduling of the WiMAX basestation is not possible. However, the physical basestation can provide congestion information through SNMP. Using this information, VNTS controls the downlink traffic of each vBTS at the network level. The VNTS controller is located on the modified vBTS ASN gateway presented in the previous subsection and monitors the channel conditions of each slice, notably the current throughput of each vBTS. The VNTS controller makes sure that the sum of the loads from each slice does not cause depletion of the

Fig. 5.7 VNTS downlink
traffic shaping mechanism
(based on [12])

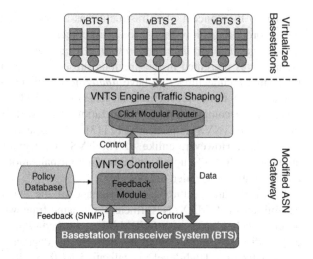

OFDMA radio resources on the physical basestation [12]. The load of each vBTS
is scaled up and down according to the weight assigned to each slice while maintaining the total load within the capacity of the basestation. The VNTS controller sends control signals to the VNTS engine which reinforces the decisions of
the algorithm at the controller. The VNTS engine is implemented using the Click
modular router [16], which takes care of the traffic shaping and routing of the
flows.

This architecture is illustrated in Fig. 5.7. Of course, true slice isolation is not
achieved due to a feedback delay and the update rate of the VNTS system. From
a theoretical perspective, this concept is similar to the principle that fairness and
optimization can be achieved when the allocation algorithm (basestation MAC
scheduler) is joined with a proper traffic shaping and congestion control algorithm
(VNTS). VNTS suggests that a virtualized wireless infrastructure can have a traffic and congestion control module at the *network gateway level* in addition to the
resource allocation module at the *air interface level*. Nevertheless, a lot of the
issues with slice isolation of vBTS and VNTS can be solved by implementing the
virtualization layer within the basestation, as will be shown in the next subsection.

5.5.3 WiMAX Virtualization: Network Virtualization Substrate

New business models, such as corporate bundle plans and services with leased networks (SLN), have been proposed in [13]. SLNs can allow a service provider to
control the quality-of-service (QoS) of their traffic over the leased infrastructure,

similar to the concept of infrastructure-as-a-service (IaaS) from cloud comput-ing. In order to achieve this, three main objectives are explored in [13]: *slice cus-tomization*, *slice isolation* and *resource usage efficiency*. The vBTS and VNTS architectures presented in the previous subsections are high-level *flow-based* vir-tualization architectures that have *limited feedback* information from the basesta-tion and no control over the internal scheduler of the basestation. To meet all three objectives, a complementary virtualization framework called network virtualiza-tion substrate (NVS) is presented in [13]. NVS is also a *flow-based* virtualization architecture. However, unlike vBTS, NVS is integrated *within* the basestation architecture. In NVS, uplink and downlink virtualization, as well as full control of the basestation scheduler, are possible.

The NVS architecture takes advantage of the existing scheduling and QoS capabilities of WiMAX. Nevertheless, the same framework can also be applied to wireless technologies with OFDMA-based scheduling, such as LTE. The authors of [13] discuss different trade-offs between the degrees of isolation and the scope of virtualization. High-level virtualization (with a wider scope) can achieve net-work-wide efficiency due to the meta-information available. However, it provides poor isolation and limited customization over scheduling algorithms. On the other hand, low-level virtualization technologies usually have much better isola-tion albeit a localized scope. In NVS, the scope of virtualization is chosen at the *flow-level* due to the fact that it can still provide good customization and research innovation while not over-exposing low-level complexity to the MVNOs [13]. Additionally, flow-based scheduling is already present in most modern cellular technologies including both WiMAX and LTE.

A distinction between *slice scheduling* and *flow scheduling* is made in [13]. These two types of scheduling represent two different scopes and can act sepa-rately from each other. Thus, each packet flow is tagged by a *slice ID* in addition to its *flow ID* [13]. Slice scheduling is the partitioning of resources across differ-ent virtual networks. The NVS framework supports two different slice scheduling policies: the *resource-based allocation* and the *bandwidth-based allocation*. The *resource-based* allocation consists of partitioning a specific proportion of OFDMA slots to a virtual network during each frame time. Even though resource-based allocation cannot guarantee the available capacity of the transmission channel at a given time, it allows the virtualized basestation to behave like its non-virtualized counterpart. The available resource limits are always known, making the virtual basestation similar to a normal basestation but with reduced data capacity. The second *bandwidth-based* approach consists of guaranteeing a specific throughput for the data flow. Here, the term 'bandwidth' refers to the *aggregate throughput* of the network and not to the frequency band. This approach allows the sched-uler to dynamically allocate resource slots to satisfy a given minimum through-put requirement. Both approaches are supported in NVS in order to cover a larger range of applications.

Flow scheduling allows MVNOs to control and customize the scheduling of downlink and uplink flows *within* their slice. Three different modes of flow sched-uling with various degrees of freedom are supported in [13]: *scheduler selection*,

Fig. 5.8 Network
virtualization substrate
framework (based on [13])

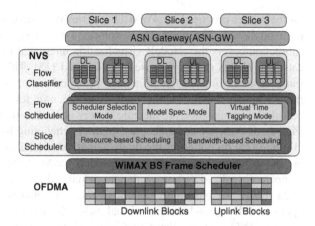

model specification and *virtual time tagging*. The *scheduler selection mode* is the
equivalent of providing a library of pre-programmed schedulers from which the
MVNOs can select from. These scheduling modules are fully autonomous and
require little or no user input. The *model specification mode* allows the MVNOs
more freedom at changing the configuration of the virtual scheduler through the
model interface and *parameters*. For example, the weighting function over each
QoS class of flows can be customized. Finally, the *virtual time tagging mode*
allows the emulation of arbitrary flow schedulers defined by the MVNOs. A virtual system time is tagged on each incoming flow in order to allow the maximum
degree of freedom in the scheduling decisions of the virtual schedulers. The overall framework is shown in Fig. 5.8.

A prototype has been implemented on a WiMAX basestation from PicoChip in
[13]. According to [13], the NVS framework is lightweight enough to perform the
scheduling of each frame within 5 ms. Finally, it is interesting to note that NVS
and vBTS are not mutually exclusive. Even though NVS has a better isolation and
scheduling mechanism, it still requires a good management interface for each virtual network operator to control their own slice. The vBTS instance can be used as
a *frontend C&M interface* that duplicates all basestation functionalities whereas
the NVS can provide *backend support* to reinforce the rules and policies configured through vBTS. In order to fully benefit from both uplink and downlink virtualization, vBTS can also be implemented *within* the basestation as opposed to
external VMs [13].

5.5.4 LTE Evolved Node B Virtualization

The 4WARD program funded through FP7 surveyed different emerging networking technologies, including the virtualization of the cellular network [14]. Based

on the model proposed by 4WARD, a series of papers on LTE virtualization has been published. The paper [15] explores a LTE virtualization framework. This subsection covers the eNodeB virtualization integrated in the scheduling of physical radio blocks (PRB) proposed by [14] and [15]. This architecture follows the same virtualization design principles previously examined in Sect. 5.3.2, using radio resource blocks as the basic resource unit for allocation. The proposed downlink virtualization architecture is very similar to that of previous subsections since both WiMAX and LTE use OFDMA scheduling in the downlink. However, [14] and [15] do not cover the implementation of LTE virtualization. Instead, they present different resource allocation models for MVNOs and simulation models to compare virtualized systems with non-virtualized systems.

In 4WARD Consortium [14], a distinction between regular mobile network operator (MNO) and mobile virtual network operator (MVNO) is made. The current MNO system allows multiple mobile operators to service in the same geographic region using different *fixed* frequency bands. In contrast, a fully-virtualized infrastructure allows multiple MVNOs to *dynamically share* the same frequency band. In fact, the term MVNO is currently used to loosely refer to service providers even without the involvement of any virtualization technologies [15]. However, these service providers have very limited configuration and control over the underlying infrastructure and scheduling of their slice. For example, *network sharing* already exists in 3G cellular networks without virtualization [5]. However, a common service agreement must be negotiated among all the providers, reducing the flexibility of the infrastructure.

Four general categories of service-level agreement (SLA) contracts are explained in [14]: fixed guarantee, dynamic guarantee with maximum bandwidth restriction, best effort with minimum guarantee and best effort with no guarantee. They are similar to the properties of flow-based QoS classes defined in the LTE standard. However, instead of being applied on individual flows, they are applied on the *aggregated flow* of the entire slice. In [14], the bandwidth requirement is translated into the number of PRB blocks reserved for each slice. It is not equivalent to the *bandwidth-based allocation* model of WiMAX NVS from the previous subsection. Instead, it is similar to the *resource-based allocation* model of NVS. This is because, unlike NVS, [14] treats bandwidth and resource as the same. The term 'bandwidth' refers to the spectrum and not to the aggregate *throughput* of the system, as in [13]. Based on the OPNET simulation in [14], the delay and throughput of non-virtualized MNOs are compared with those of MVNOs. In the MNO case, only a fixed-guarantee contract can be implemented whereas in the MVNO case, all four types of contracts can co-exist. The overall delay and throughput performance of the system is similar in both cases for fixed and dynamic guarantees. However, the MVNO system has the advantage that the left-over bandwidth from a high-tier contract can be harnessed by a lower-tier best-effort contract, improving the usage efficiency of the overall system.

The LTE virtualization framework proposed in [15] focuses on the LTE evolved Node-B (eNodeB or eNB). The eNBs are the equivalent of basestations in LTE and play the role of scheduling and allocation of the radio resources to client users. Even

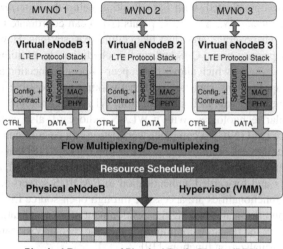

Fig. 5.9 LTE eNodeB downlink virtualization (based on [15])

without virtualization, the eNB already contains default scheduling capabilities using multiple access techniques such as OFDMA. However, in order to achieve virtualization, additional inter-slice isolation and allocation mechanisms must be integrated within the existing scheduler [15]. In general, based on different criteria such as throughput, power and channel conditions, the number of PRBs allocated to each virtual network at any given time can be derived. As shown in Fig. 5.9, the general LTE virtualization architecture is similar to the configurable radio network architecture presented in Sect. 5.3.2. The hypervisor layer takes the configuration and contract requirements of each virtual eNB and dynamically allocates the PRBs to each slice. This hypervisor can operate as a virtualization agent or resource broker.

Overall, the proposed LTE virtualization framework is extremely similar to the other cellular network virtualization architectures presented in this section, especially NVS for WiMAX. Even though [14] and [15] only presents the downlink virtualization, uplink virtualization for LTE is covered in [17] by the same research group. Similar to NVS, the LTE virtualization architecture assumes a full and open access to the underlying MAC layer of the basestation. Thus, the implementation is not possible unless the virtualization architecture is integrated as a new feature by the equipment manufacturers, as in the case of NVS with NEC [13].

5.6 Virtualization in 802.11 Wireless Local Area Networks

Compared to cellular technologies, IEEE 802.11 wireless local area networks (WLAN) have the advantage of supporting faster and easier deployment (concept of plug-and-play). However, as a trade-off, the control and management framework is relatively less sophisticated, with the exception being the *enterprise WLAN*. In the *infrastructure mode*, WLAN access points (APs) can coordinate multiple client devices within their coverage area, just like cellular basestations. However, there is only support for priority-based QoS but not for scheduling. In addition, 802.11 APs support the *ad hoc mode*, which allows direct peer-to-peer connection among 802.11 devices. Thus, applications such as mesh networking are possible using 802.11-based technologies. In terms of implementation, WLAN technologies are more versatile than cellular technologies. With a WLAN radio card, also called a wireless network interface card (NIC), a computer workstation can be configured to act as an 802.11 AP. Virtual access point (VAP) is an existing concept [18] that can be implemented with features such as multiple service set identifiers (SSIDs), association to different VLANs and forwarded authentication to RADIUS servers. Overall, many of these 802.11 functionalities are accessible through open-source Linux-based software packages and firmware, making the development of a virtualization framework more accessible for the research community. Thus, the regular VAP can be enhanced with additional virtualization functions [19].

In this section, different virtualization architectures and implementations applied on 802.11 WLAN technologies are presented. First, SplitAP [20], a virtualization framework split across the AP and the client device, is covered. This is followed by the presentation of a wireless NIC virtualization architecture [21] for VMs, similar to the Ethernet NIC virtualization in Sect. 4.3 of Chap. 4. Then, another wireless NIC virtualization architecture based on existing MAC functions such as the point coordination function (PCF) and the power-saving mode (PSM) [22] is discussed. Finally, the virtualization of a *heterogeneous mesh network* with both 802.11 WLAN nodes and 802.16 WiMAX nodes using a MAC abstraction layer [23] is overviewed.

5.6.1 SplitAP Architecture

This subsection covers the SplitAP architecture for 802.11 interfaces proposed in [20]. Whereas the VNTS presented in Sect. 5.5.2 is useful at providing fairness in *downlink* traffic, a different approach must be applied when handling *uplink* traffic. This is especially important in the context of 802.11 WLAN technologies, which lack a *dedicated control channel* between the AP and the wireless client stations. In Bhanage et al. [20], the SplitAP architecture is aimed at providing a fair allocation of air-time resources in the uplink transmission. This architecture is spread across both the client station and the access point. The main contribution of SplitAP is the addition of a *feedback mechanism* between the client station and the access point in order to schedule the uplink resources.

SplitAP is divided into two components: the *SplitAP controller* and the *client plug-in*. The SplitAP controller is added as an extension to the virtual access point (VAP) functionalities discussed in [18]. The VAP advertises a different SSID for each *virtual WLAN*, giving the illusion of multiple access points over the same physical hardware. Each VAP is able to provide a distinct set of management functionalities such as different security configurations, VLAN association and virtual SNMP management information base (MIB). In [18], virtual SSIDs are suggested to be implemented with *multiple beacon advertisements* each with a distinct basic service set identifier (BSSID), which is a MAC address, and a distinct extended service set identifier (ESSID), which is a string representing the WLAN name. Most modern enterprise access points support these VAP functionalities [20]. The access point in SplitAP heavily leverages from these existing functionalities.

At the access point, the SplitAP controller introduces an uplink scheduler with two scheduling algorithms: the linear proportional feedback control (LPFC) and a modified version of LPFC (LPFC+) proposed by [20]. From the perspective of the controller, the clients are aggregated into user groups assigned to different virtual WLAN. However, the SplitAP architecture is based on commercial 802.11a wireless chipsets with limited access to PHY layer functionalities. Thus, SplitAP can only time-share resources. In the basic LPFC algorithm, each virtual WLAN is allocated a slice of the total uplink air-time resources. Then, within each slice, the algorithm proportionally allocates uplink air-time resources based on the current number of connected clients in that VAP. On the other hand, the modified LPFC+ algorithm dynamically measures the actual utilization of the channel by each client and adjusts their quota accordingly. The client-side plugin software is an application installed as a separate module to exchange control messages with the SplitAP controller. It performs *outbound traffic shaping* on the client, implemented using the Click modular router. The overall architecture is shown in Fig. 5.10.

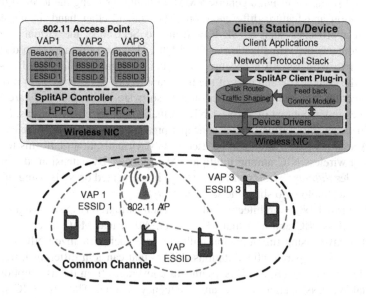

Fig. 5.10 SplitAP controller and client plug-in with virtual access points (based on [18] and [20])

Without the SplitAP controller and plug-in, the uplink air-time resource consumption is a competitive model entirely decided by the throughput of the individual clients. With LPFC, a fixed amount of air-time is always *guaranteed* for each client. On the other hand, LPFC+ allows a *dynamic sharing* of air-time resources among the clients and across the slices. These benefits are similar to those achieved through LTE virtualization discussed in [14] and Sect. 5.5.4. Overall, [20] provides a way to control the client uplink traffic from the network layer through a client plug-in, which adds a dedicated *application-level* control channel between the AP/controller and the client stations.

5.6.2 Wireless Local Area Network Client Interface Virtualization

According to [21], the degree of freedom offered by existing Ethernet NIC virtualization software in host computer virtualization platforms is not sufficient for wireless interfaces. For instance, functions such as user authentication and user mobility do not have equivalent representations in Ethernet interfaces. The consideration for a set of additional functions such as rate adaptation and power management makes WLAN virtualization quite different from Ethernet LAN virtualization. Existing Ethernet NIC implementations are mainly classified as software-based or hardware-based. On one hand, in the software-based approach, the virtual machine monitor (VMM) or hypervisor implements the de-multiplexing of different virtual NICs. Due to the large variety of features from different NIC vendors, it is difficult to design a generic VMM without sacrificing the level of device customization and feature differentiation [21]. On the other hand, in hardware-enhanced approaches such as SR-IOV (refer to Sect. 4.3), the virtual machine directly contacts the virtualization-enabled NIC. However, this approach does not scale well to wireless NICs because additional management functionalities will increase the complexity and cost of the hardware device [21]. Thus, [21] proposes an architecture that is a combination of *both* software-based and hardware-based approaches in the context of wireless NIC virtualization for client stations. Similar to the Ethernet NIC virtualization concepts presented in Sect. 4.3, wireless NIC or WLAN card virtualization can be used in host virtualization platforms to share a physical wireless NIC among multiple VM instances. By extension, this potentially enables *different wireless firmware* to be implemented on VMs using this virtualization technology to share the *same radio card*.

In the virtual Wi-Fi architecture of [21], each virtual WLAN is assigned to a virtual wireless NIC and a virtual MAC layer. All virtual MAC layers share the same PHY layer using time-division multiplexing. Each virtualized NIC instance can operate as a separate client stations (STAs) that can join different infrastructure basic service sets (BSS). This architecture leverages from the assumption that all virtual WLANs operate on the same frequency channel. Thus, the NIC simply receives packets destined for *all* virtual WLANs and filters those based on MAC addresses to match with their respective virtual MAC layer. The use of different frequency channels and virtual PHY layers for different virtual MAC layers is outside the scope of [21] but can be achieved through SDR technologies and spectrum virtualization techniques respectively discussed in Sects. 5.7 and 5.8.

Shown in Fig. 5.11, the virtual Wi-Fi architecture is divided into four main components: the *guest machine wireless NIC driver*, the virtual Wi-Fi *device model*, the *virtualization-augmented device driver* and the *virtualization-augmented NIC*. This hybrid architecture borrows the hardware-based concepts of SR-IOV but keeps a software-based interface abstraction at the VMM layer. The guest machine wireless NIC driver has the same functionality as the native driver and is provided by the NIC manufacturer. Through a generic device model interface which only implements basic peripheral component interconnect (PCI) I/O functions [21], the guest driver communicates with the virtualization-augmented device driver residing in the host machine. As a new business model, [21] suggests that these augmented drivers should be provided by the NIC manufacturers, which allows them to offer competitive *differentiated features*. Similar to the SR-IOV discussed in Sect. 4.3, each virtual instance should have a *dedicated* software and hardware lane, maintaining a certain degree of isolation across the layers. The augmented driver de-multiplexes the management control messages obtained from the virtual drivers and passes them to their corresponding virtual MAC layer located on the augmented NIC. The microcontroller code of the augmented NIC is modified to accommodate multiple virtual instances by keeping the state of each slice isolated [21].

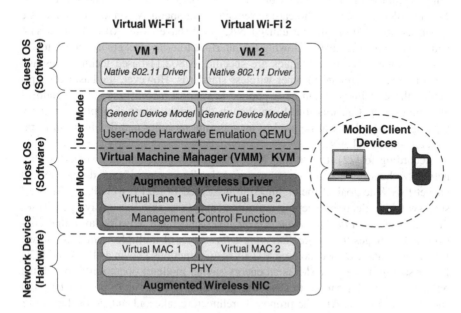

Fig. 5.11 Virtual wireless NIC architecture (based on [21])

In [21], the virtual machines are implemented using KVM with QEMU. The virtual device model is implemented in QEMU and handles the creation of new virtual Wi-Fi instances. The augmented device driver on the host is a modified version of the Intel Wireless Wi-Fi driver part of the Linux kernel. The augmented NIC is an extension to the Intel My WiFi technology which is able to support two virtual MAC layers. Interrupt coalescing and hardware-based address translation are implemented to reduce the CPU virtualization overhead [21]. The performance evaluation in [21] indicates that their virtual NIC has 15 % extra delay compared to the native NIC. According to [21], some wireless management functions such as power management can be implemented using PSM. Nevertheless, some functions such as transmission power remain hard to share due to conflicting requirements. Specific workarounds must be applied to resolve conflicts in the configurations between virtual instances. Overall, the main advantage of this architecture is that it can enable the usage of *native* wireless drivers inside the guest operating systems. However, as disadvantage, it requires modifications and integration across all existing components.

5.6.3 802.11 PSM and PCF Virtualization for Mesh Network

This subsection presents the virtualization of wireless NICs, applied in mesh networks, based on the 802.11 MAC functions such as the point coordination function (PCF) and the power-saving mode (PSM) [22]. The authors of [22] present an implementation of virtualization based on manipulating the PCF and PSM of the NIC in order to allow a radio card to remain simultaneously connected to multiple networks. In addition, the virtualized NIC can behave as an AP and client STA simultaneously, effectively allowing the interface to relay packets in a mesh topology. Of course, these techniques are specific to 802.11-based technologies and cannot be easily applied to other wireless standards. However, the basic concepts behind these techniques can be re-applied to support mobility, seamless handover and multi-point transmission. Even though this architecture is applied on mesh topologies in [22], it can be used to increase the mobility and flexibility in infrastructure-mode WLANs.

According to [22], the current NIC must disassociate and re-associate with a network whenever a handover is required, leading to significant downtime and packet loss. The goal of the architecture presented in [22] is to reduce and avoid such switching delay by keeping multiple connections simultaneously open. Thus, there is a *virtual interface management layer* under the network layer that controls and manages the *connection state information* of multiple virtual interfaces. It takes advantage of two existing 802.11 MAC functions known as PSM and PCF. Power saving mode, or PSM, is an energy-saving mode of operation in which the wireless client STA can go to sleep and periodically wake up to poll for buffered frames stored in the AP. The proposed architecture takes advantage of this buffer in order to prevent the loss of packets during the switching between two virtual networks. For instance, when the client-mode NIC must switch connection, a *PSM doze signal* is sent to the AP belonging to the initial network.

The point coordination function, or PCF, is an optional MAC mode for infrastructure-mode WLAN that has priority over the more commonly-used distributed coordination function, or DCF. The MAC is centralized in PCF at the AP instead of being distributed as in DCF. Virtualization is hard to achieve using DCF due to the distributed nature of the back-off algorithms. In PCF, the AP acts as a master coordinator that polls the client STAs for transmission frames. The virtualization architecture in [22] modified PCF to become virtualization-aware, giving the virtualized station in AP mode the centralized control over which associated *virtual* client STAs to poll. In a more advanced infrastructure requiring QoS, 802.11e introduces the hybrid coordination function (HCF), four different QoS *access categories*, as well as an enhanced PSM called the automatic power-save delivery (APSD) [1]. The enhanced distributed channel access (EDCA) replaces DCF and the HCF controlled channel access (HCCA) replaces PCF. Unfortunately, [22] does not address these more advanced features.

A mesh network topology that contains these virtualized interfaces is shown in Fig. 5.12. As one of the possible applications illustrated in the figure, the infrastructure-mode WLAN, similar to a cell in the cellular network, is *extended* by a mesh network through a *virtualized node*. These virtualized wireless nodes contain a virtualized NIC that is used to reach fringe client STAs that are not covered by the existing APs.

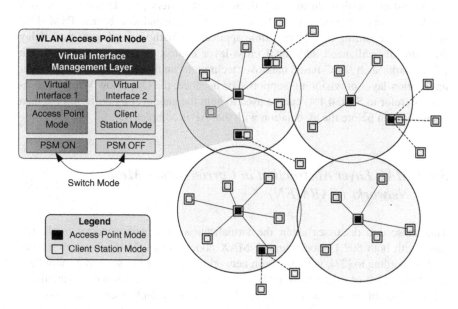

Fig. 5.12   Virtualized wireless NIC for mesh network using PSM and PCF (based on [22])

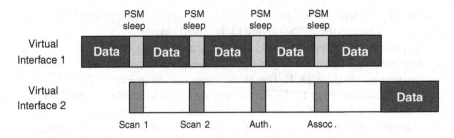

Fig. 5.13 Soft handover based on virtual interfaces using PSM (based on [22])

Other than mesh networking applications, [22] also discusses the possibility of
using the virtualized interface to perform seamless soft handovers. In this application, the wireless NIC of a client STA can perform scanning, authentication
and association with a second AP without disassociating from the first AP. The
handover protocol defined in [22] consists of incrementally performing the steps
required for association. While scanning and authenticating with a second access
point, the NIC will constantly switch back to the original connection to fetch
any packets that arrived in the buffer during the PSM doze period, as shown in
Fig. 5.13. This guarantees that delay-sensitive applications are still serviced within
an acceptable time frame.

Unfortunately, from the simulation results shown in [22], these techniques have
a limited scalability in terms of the supported number of virtual networks. As the
number of simultaneous connections increases, the PSM sleep time for each connection also increases, leading to higher packet delays and potential packet loss
due to buffer overflow during high-throughput transfers [22]. In the case of soft
handover, worst case delays of about 10-110 ms are induced by the PSM doze
state [22]. According to [22], real-time applications with very stringent time delay
requirements will need additional cross-layer control to reduce the latency. In
other words, such MAC-based handover technique can be coupled with network or
application layer provisioning support. For instance, the OpenFlow Wireless technology (refer to Sect. 4.1.4) can be used to help the handover by buffering the data
at an AP even before the association with that AP is completed.

5.6.4 MAC Layer Abstraction in Carrier-Grade Mesh
Networks (CARMEN)

This subsection discusses about the virtualization of a heterogeneous mesh network with both 802.11 nodes and WiMAX nodes using a MAC abstraction layer
[23]. According to [23], wireless mesh networks can be used to support a more flexible deployment of the network infrastructure in non-urban environment. Intelligent
self-configurable mesh networks are ideal for temporary deployment. The mesh networking research project discussed in [23], called *carrier grade mesh networks* or

CARMEN (not to be confused by the cloud-centric architecture Carmen presented in Sect. 5.4.1), is initiated by the European Seventh Framework Programmes (FP7). However, the control and management of such network can be difficult, especially with the inclusion of heterogeneous wireless access technologies in order to benefit from the diversity of the coverage. Thus, [23] proposes a MAC abstraction layer to combine both 802.11 and WiMAX nodes in the same carrier mesh network.

The CARMEN mesh network is composed of three different types of nodes: CARMEN mesh points (CMPs), CARMEN access points (CAPs) and CARMEN gateways (CGWs). The CMP nodes serve as relays to other nodes and forwards packets coming from CAPs with *flow-based* quality-of-service (QoS) reinforcements like a wireless router. The CAP nodes are located at the edge of the mesh network and directly service the end users. The CGW nodes are the boundary points between the wired network infrastructure and the mesh network. The CARMEN nodes and an example network topology are shown in Fig. 5.14.

The MAC abstraction layer proposed in [23] has the objective of simplifying the management and traffic offloading of the mesh nodes. It is divided into two sub-layers: the *mesh functions sub-layer* and the *MAC abstraction sub-layer*, as shown in Fig. 5.14. The mesh functions sub-layer consists of the *technology-independent* mesh node management functions, including mobility, forwarding, monitoring and configuration. These functions are separated in modules that can be located on the node or remotely controlled [23]. These functions interface with the *technology-independent* part of the abstraction layer. The interface management function (IMF) maps the mesh functions from the technology-independent interface to the *technology-dependent* interface through a *MAC adapter*.

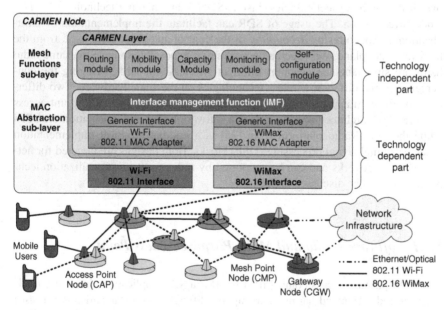

Fig. 5.14 MAC layer abstraction and carrier grade mesh network architecture (based on [23])

The abstraction layer is applied on both the *data plane* and the *control plane*. The data plane is composed of aggregated flows identified with a *pipe ID* [23]. The routing of the pipes can be performed based on MPLS, IPv6 flow labels or OpenFlow (refer to Sect. 4.1). The abstraction layer provides the interface necessary to control the resource allocation and the QoS configuration of each pipe. The control plane is integrated with the abstraction layer through an extension of the IEEE 802.21 handover and interoperability standard to support self-configuration and spectrum management [23]. Overall, [23] is based on a *modular* architecture and the abstraction of MAC functions of different access technologies. Nevertheless, while it provides a framework to manage heterogeneous nodes, CARMEN does not virtualize the underlying radio interfaces. In other words, multiple RF frontends must be attached on the same node if multiple access technologies are to be used. Then again, this abstraction layer can be combined with advanced SDR technologies and spectrum virtualization techniques that will be presented in the following sections in order to support a fully-virtualized infrastructure.

5.7 Virtualization and Software-Defined Radio

The goal of software-defined radio (SDR) technologies is to enable a more programmable and general-purpose radio hardware, similar to the effects of software-defined-networking (SDN) technologies on the network fabric. By itself, SDR does not necessarily imply wireless virtualization. Instead, it can be viewed as one of the enabling technologies, just as SDN is an enabling technology for network virtualization. The usage of SDR can facilitate the implementation of virtualization architectures by *decoupling* of the control and processing logic from the hardware data plane. With such a decoupling, it is possible to have the same radio hardware simultaneously support different wireless standards, bringing the convergence of the different wireless technologies on the same hardware. Two different virtualization approaches are presented in this section. The multi-point access point (MPAP) architecture from [24] is a software-based implementation on GPPs while the OpenRadio platform from [25] is a middleware-level implementation based on digital signal processing (DSP) blocks. Both designs are targeted for heterogeneous networks and can be supported by radio spectrum virtualization techniques that will be discussed in Sect. 5.8.

5.7.1 Software Radio and Multi-Purpose Access Point

The multi-purpose access point (MPAP) [24] is a SDR application aimed at virtualizing the radio frontend hardware to support different radio standards. MPAP uses the Software Radio (Sora) platform developed by Microsoft Research Asia [26].

The main advantage of MPAP is that it enables the same radio frontend to be used for 802.11g and 802.15.4 (ZigBee) with good real-time performance, while remaining entirely *software-based*.

The Sora platform is designed to solve many of the shortcomings of using GPPs to execute operations traditionally performed in DSP blocks or field-programmable gate arrays (FPGAs). According to [26], while SDR technologies based on programmable hardware using DSP and FPGA are very fast, they are often more expensive and do not provide the same amount of flexibility as a purely software-based development platform. On the other hand, without proper optimization, it is very hard for GPPs to meet the throughput and latency requirements for practical real-time radio transmission. The Sora platform presented in [26] attempts to address these issues. Sora is divided into three main sections: the *radio front-end* hardware, the *radio control board* (RCB) and the *software radio stack*.

The radio front-end hardware consists of the antenna and the digital/analog converters. Currently, the Sora platform can use the Universal Software Radio Peripheral (USRP) or the Wireless Open-Access Research Platform (WARP) as radio front-end [26]. The main feature of Sora is that the digitized signal is directly fed into a host computer through the RCB. Thus, the entire PHY layer baseband processing is performed in software. The main challenges of baseband processing in software are the bus throughput and the latency associated with the transfer of large amount of sampling data. According to [26], 802.11a/b/g requires a 1.2 Gbps bus throughput for each 20 MHz channel and the 802.11 MAC can only tolerate acknowledgement (ACK) delays of around 10 ms. The RCB is an interface board that solves these issues by using a Xilinx Virtex®-5 FPGA to implement a direct memory access (DMA) module and a custom PCI Express (PCIe) controller, which supports a bus transfer rate up to 16.8 Gbps and a latency of only hundreds of nanoseconds [26], which can satisfy many of the 802.11 timing requirements. The RCB can be installed on any computer workstation like a regular network interface card (NIC). Nevertheless, additional software-based optimizations are still required in order to provide real-time PHY layer processing.

The software radio stack, which resides in the operating system kernel as a driver-like component, contains a library of PHY functions optimized for multicore GPPs using streamline processing, look-up tables (LUTs) and single instruction multiple data (SIMD). The LUT optimization consists of converting PHY processing tasks into LUTs that can reside in the L2 cache of processor cores. According to [26], this improves the software processing speed by a factor of 1.5–50. The SIMD technique consists of using 128-bits vector instructions originally designed for graphic cards to perform parallel processing of sampled data within a *single instruction*. The software components are run on *dedicated* processing cores in order to satisfy real-time requirements. In [26], *SoftWifi* and *SoftLTE* are implemented as examples of software implementation of the standards. SoftLTE, a software-based LTE uplink shared channel (PHUSC), can reach a data rate of 43.8 Mbps for a 20 MHz channel with 16QAM modulation and ¾ Turbo coding while using only a *single* hyper-threaded Intel i7-920 core. On the other hand, SoftWifi can achieve full data rate with 802.11g and is fully inter-operational with regular commercial Wi-Fi devices. In future releases, the Sora platform will also support MIMO transmission [26]. The overall Sora architecture with MPAP is shown in Fig. 5.15.

The MPAP architecture presented in [24] is one possible application of Sora and consists of adding virtual NIC (VNIC) instances over the Sora software radio stack. A new layer called the *SDR service layer* acts as the hypervisor of the architecture. It contains a scheduler to minimize the interference between the different virtual instances running over the same radio front-end. A demonstration is shown in [24] with the coexistence of two ZigBee VNICs and one 802.11g VNIC. As mentioned previously, one main advantage of virtualization is the *opportunistic sharing* of bandwidth resources. MPAP demonstrates this by using OFDM on *non-contiguous* bands for the 802.11g VNIC in order to allow narrow-band ZigBee to share the same band. Once again, the contribution of the MPAP and Sora is that the entire architecture runs as software on GPPs while satisfying the real-time requirements of the protocols. Alternative fully software-based SDR platforms include the GNU Radio [27] and [28]. Although GNU Radio suffers from real-time performance limitations, it has a richer set of open-source components developed by various research groups. PCIe interface cards can also be used in GNU Radio to remove some of its hardware limitations. This is contrasted with Sora, which has a much better performance but a relatively

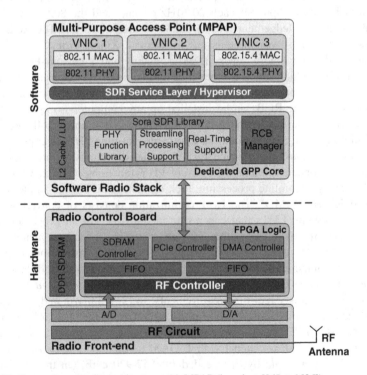

Fig. 5.15 Sora software radio architecture with MPAP (based on [24] and [26])

new penetration in research communities [29]. A spectrum virtualization layer (SVL), which will be covered in more detail in Sect. 5.8.1, can be implemented on Sora.

5.7.2 OpenRadio Programmable Wireless Dataplane

OpenRadio is a SDR platform under development by Stanford University proposed in [25]. Once again, OpenRadio in itself is not a virtualization framework, but a programmable framework to implement the open-infrastructure concept inspired by technologies such as OpenFlow (refer to Sect. 4.1). Whereas the Sora platform pushes the entire SDR architecture into software, the OpenRadio platform keeps the baseband processing implemented in programmable DSP blocks. The goal of OpenRadio is to provide *configurability* and *programmability* of PHY-layer protocols. The authors of [25] justify the need of such SDR technologies due to the accelerated *continuous evolution* of wireless protocols and the *densification* of wireless networks. Thus, the usage of a flexible and generic wireless radio dataplane will reduce both the deployment time and cost of constantly upgrading the wireless infrastructure [25]. OpenRadio presents two major contributions: the decoupling of wireless protocols from hardware and the separation of *processing* and *decision* functions in wireless protocols.

Similar to how OpenFlow enables the programmability of switches and routers through the OpenFlow API, OpenRadio allows the virtual operator to use a declarative *rule-action*-based programming style to define different wireless protocols. First, the decoupling of wireless protocols from the physical hardware is performed through *protocol decomposition*. In other words, a set of basic irreducible PHY and MAC building blocks are identified for a series of common wireless protocols. Each of these building blocks is associated with a particular piece of hardware. PHY building blocks, such as fast Fourier transform (FFT), channel estimation and forward error correction (FEC) [25], are all implemented within DSP and FPGA-based modules. Since these functions are common to most wireless PHY layers and only differ in parameter settings, a *common dataplane* can be constructed for most if not all wireless protocols. Thus, the wireless protocol becomes a *logical assembly* of basic functional blocks. The protocol itself consists of two distinct planes: the *decision plane* and the *processing plane*. The decision plane is a logic assembly of *rules* and *decision branching* points [25]. The processing plane is defined by *actions*, which are mathematical operations over the digitized data. The decision plane and the processing plane of OpenRadio are separated by the *information plane*, an interface formed by the configuration and statistics of the processing plane acting as interface between the two planes [25]. Using this control model, OpenRadio can allow *application-specific* and *flow-specific* treatment of data at the MAC and PHY level.

The prototype implementation of OpenRadio is a hybrid software-hardware architecture based on the Texas Instrument KeyStone™ DSP platform [30]. According to [25], only variants of the 802.11 protocol were chosen for implementation to demonstrate the OpenRadio concept. However, the same under-lying architecture can be extended to any wireless standards. Similar to the chal-lenges presented in Sora, the most stringent timing requirement of 802.11 comes from the ACK reply time of 25 μs [25]. In order to meet these deadlines, high efficiency of the architecture is maintained by limiting the scope of the flexibility of building blocks to the bare minimum required to implement a given wireless protocol. Additionally, the rule-action programming style and the decoupling of the decision and processing plane allow a *fixed guarantee* on the execution time of functions [25]. This is because the time-intensive data processing is performed in hardware (fixed execution time) while all the less computationally-intensive decision branching are performed in software (variable execution time). This decoupling allows OpenRadio to near-deterministically meet real-time timing requirements [25]. This is one main advantage of OpenRadio over pure software-based SDR platforms such as Sora. In order to increase deterministic behavior of overall system, the software decision plane follows a *static scheduling* with *event polling* as opposed to *event interrupt*. A flow-based finite-state-machine (FSM) approach is applied on the decision plane. A summary of the architecture is shown in Fig. 5.16.

The first OpenRadio prototype is able to achieve 54 Mbps in 802.11g mode with approximately 20 % of the computational load caused by overhead in the architecture [25]. The main limitation of OpenRadio is the assumption that the computational intensity of the decision plane is relatively negligible compared to that of the processing plane. Otherwise, the overhead in the decision plane will be too large, making the system inefficient [25]. Similar to Sora, the OpenRadio plat-form can be used to virtualize MAC- and PHY-layer resources. Ultimately, it can interoperate with SDN technologies such as OpenFlow in order to provide a fully

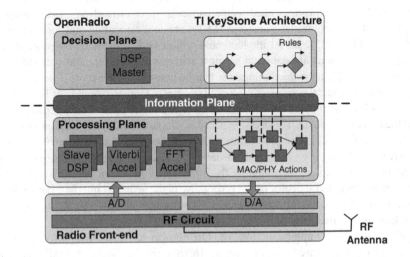

Fig. 5.16 OpenRadio architecture on TI KeyStone platform (based on [25])

configurable wireless network infrastructure [25]. The radio front-end can also be enhanced by using a radio slicing technology such as Picasso, which will be presented in Sect. 5.8.2.

5.8 Virtualization by Spectrum and Radio Frontend Slicing

In order to allow the efficient reuse of the frequency spectrum, the spectrum can be abstracted and dynamically allocated to different radio protocols, which can be facilitated through *spectrum virtualization*. On the other hand, in *RF frontend virtualization* and *radio slicing*, the RF frontend circuit itself can be enhanced to support full duplex and wideband transmission with high degree of RF isolation between slices. In this section, spectrum virtualization and radio frontend virtualization techniques are explored, more particularly the spectrum virtualization layer (SVL) of [31] and the Picasso architecture of [32]. Spectrum and RF frontend slicing are the lowest levels of resource slicing possible and allow the highest degree of flexibility and granularity. Both are potentially important enabling technologies for virtualization in *dense* and *heterogeneous* wireless networks, due to the stringent requirements in spectrum sharing and management. These techniques can complement each other and complement the SDR technologies presented in Sect. 5.7 since they are respectively developed by the same research groups.

5.8.1 Spectrum Virtualization Layer

The spectrum virtualization layer (SVL) architecture presented in [31] is designed as a sub-PHY 'layer 0.5' residing *between* the PHY layer and the RF circuit. SVL is a possible solution to achieve dynamic spectrum management and access (DSA) which can in turn facilitate emergent technologies such as *cognitive radios*. In Sect. 5.5, virtualization in cellular networks was shown to improve spectrum re-use by dynamically sharing the wireless airtime resources. However, the flexibility in channel selection is ultimately limited by the specific wireless standard. According to [31], technology-specific integration of DSA is complex, costly and not very scalable. Thus, SVL provides a *transparent abstraction layer* for the spectrum allocation, allowing the DSA to be implemented in a *technology-independent* spectrum manager. By introducing an intermediate layer, SVL does not require major modifications to the MAC and PHY layers of existing wireless standards. Not only SVL can make *non-contiguous* channels *virtually contiguous*, it can also *reshape* the channel bandwidth and shift the frequency band of the carrier. In addition, SVL also enables the abstraction of radio frontends. This allows applications such as *white-space networking* to use the 802.11g protocol to transmit data in the TV channel frequencies [31]. Even though in [31], the SVL architecture is fully implemented in software by using the Sora SDR platform discussed in Sect. 5.7.1, the same architecture can also be applied in dedicated hardware platforms, although with a reduced degree of flexibility.

The SVL architecture is composed of a few major components: the *spectrum manager*, the *spectrum map*, the *spectrum reshaper* and the *software mixer/splitter*. The authors of [31] make an analogy between SVL and routers. The spectrum manager implements a *spectrum management layer* equivalent to a 'routing protocol' to determine the spectrum allocation instead of route. The spectrum manager computes the spectrum map through allocation algorithms and policies. On the other hand, the spectrum map is the equivalent of a 'forwarding table' for spectrum. It is designed to be simple and efficient with as little overhead as possible. The spectrum reshaper performs the role of the *virtualization layer* that reshapes the baseband spectrum of each slice to fit into the physical baseband spectrum allocated through the spectrum map. Finally, the mixer and splitter multiplexes/de-multiplexes the baseband signals of each slice before sending the combined spectrum to the radio frontend. Due to the abstraction layer provided by SVL, multiple radio frontends can be managed through the same framework. The overall SVL architecture is shown in Fig. 5.17.

As shown in Fig. 5.17, the spectrum reshaper performs three basic functions: *signal decomposition and re-composition, bandwidth and sampling rate adjustment*, and frequency *shifting*. On the transmitter side, the slice baseband signal is decomposed using FFT to transform it from the time domain to the frequency domain. Then, the frequency domain signals are re-composed into potentially non-contiguous bands using IFFT. The procedure to select the FFT and IFFT size are presented in greater details in [31]. The bandwidth and sampling rate adjustments

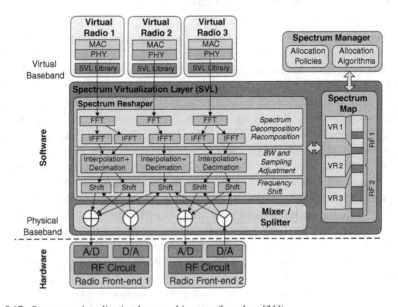

Fig. 5.17 Spectrum virtualization layer architecture (based on [31])

are performed by using basic signal processing techniques such as interpolation and decimation. The center frequencies of the allocated carriers are then shifted and mapped into the shared physical baseband. The reshaping of the virtual baseband signal can potentially introduce additional diversity and multipath effects not presented in the non-virtualized system [31]. In the case of non-contiguous mapping, the virtual baseband spectrum can suffer from different channel conditions at different frequencies, introducing *internal* fading-like effects. However, [31] argues that these effects can be reduced through the use of time synchronization sequences and that many PHY layers have strong protection against multipath fading. In the worst case scenario, the non-contiguous mapping option can be turned off for PHY layers sensible to multipath effects [31]. Finally, since the sampling rate and bandwidth are either extended or compressed, a *virtual timing system* is introduced to scale the timing of the protocols accordingly. Each virtual PHY layer uses the *virtual clock* instead of the real-time clock module. This timing system allows SVL to provide full virtualization support of any wireless protocols running at *arbitrary* frequencies. The downside of this approach is that SVL is required on both the transmitter and receiver side. For legacy connections, the virtual timing system can be turned off by preventing the rescaling of the virtual baseband spectrum.

The SVL prototype is implemented using the Sora SDR platform. In [31], each virtual PHY layer runs in a dedicated user-mode thread using the Sora SDK library and the SVL library. The spectrum manager runs as a separate process. The measured throughput of a virtual baseband mapped into non-contiguous bands is the same as the contiguous case [31]. The SVL overhead is estimated to be approximately 0.6 CPU for contiguous mapping and 0.9 CPU for non-contiguous mapping due to extra signal processing required [31]. Ultimately, the flexibility of the DSA can be limited by the duplex and wideband capabilities of the radio frontend. To solve this issue, SVL can potentially interface with RF frontend slicing techniques such as Picasso that will be discussed in the following subsection.

5.8.2 Spectrum Slicing with Picasso

The radio frontend hardware is the last piece of the puzzle missing in a fully virtualized wireless infrastructure. The frontend circuit ultimately limits the flexibility of the virtualized infrastructure. An example of such limitations discussed in [32] is that a regular RF frontend does not support full duplex transmission. This is due to the transmitting signal being orders of magnitude higher than the received signal, saturating the analog to digital converter (ADC) even if they are not using the same frequency spectrum. In 802.11, the difference in power can reach up to 80 dB [32]. Simple solutions include the use of static filters and parallel radio chains. However, these techniques are not suitable for the dynamic spectrum required for virtualization and not feasible to implement in mobile devices [32]. Thus, Picasso, a RF frontend that provides spectrum isolation through *passive self-interference cancellation* is proposed in [32]. The Picasso architecture allows simultaneous transmission and reception of signals in the same spectrum with a single antenna and reduced leakage in adjacent spectrum. In addition, it provides spectrum slicing and virtualization through FPGA-based digital filters as an alternative to the pure software-based SVL approach discussed in the previous subsection.

According to [32], Picasso allows multiple virtual wireless networks to use DSA to share fragmented spectrum in the same frequency band with a higher degree of flexibility (full duplex). For instance, Picasso can allow slices to transmit and receive at the same time, which is not possible on a regular RF frontend without support for duplex transmission. In order to achieve this, two components are used: the *passive self-interference cancellation system* and the *programmable spectrum slicing engine*. The passive self-interference cancellation system is based on a full-duplex circuit composed of a circulator and a passive cancellation circuit. In [32], the cancellation range is determined by the dynamic range of the ADC and the operational range of the wireless signals. The cancellation required for 802.11 in the worst-case scenario is at least 45 dB [32]. In Picasso, a three-port circulator with 15 dB isolation between transmit and receive ports is used. Then, using a power splitter, 15 % of the transmitted signal is redirected to the passive cancellation circuit. Due to the exclusive use of passive components, only *fixed* delay lines can be used. Since delay synchronization is extremely important in cancellation, passive programmable attenuators are used to compensate for small variations caused by temperature, wire length and external factors [32]. Two delay lines as opposed to a single line are used in order to provide the extra degree of freedom to change the phase of the reference signal. The programmable attenuators on the two delays are updated every millisecond [32]. The reference signal is then subtracted on the receiver line using a balun. A combined cancellation of 45 dB is reported in [32], which is enough to prevent ADC saturation in worse case scenarios.

The spectrum slicing engine is implemented using Xilinx Virtex-5 FPGAs and contains a digital filter engine to perform re-sampling, filtering and remapping of the digitized samples. The overall spectrum slicing architecture is comparable to that of SVL except that it is mainly FPGA-based. In the transmission chain of Picasso, the re-sampling is first performed through up-sampling and interpolation [32]. Filtering then removes the aliases generated by the up-sampling. Finally, the spectrum map determines the intermediate frequencies before the final up-conversion to the carrier frequency [32]. It is important to note that additional digital self-interference cancellation is performed inside the slicing engine to provide further isolation of the spectrum. As shown in Fig. 5.18, the spectrum slicing layer acts as a spectrum hypervisor similar to SVL that can simultaneously support multiple MAC and PHY layers on a single antenna. Experimental measurements were performed in [32] and show that the platform can support at least four virtual radio stacks without much degradation in throughput. At high data rates, there is a degradation of around 10 % due to higher signal power and leakage [32]. This degradation can be reduced by increasing the minimum frequency separation between

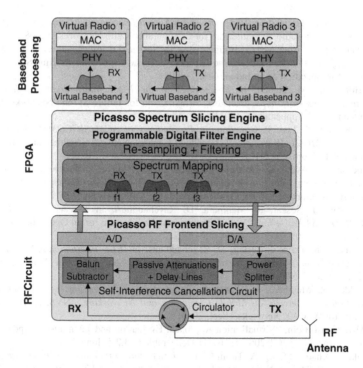

Fig. 5.18 Picasso spectrum slicing and self-interference cancellation architecture (based on [32])

slices at the cost of wasting more resources. Overall, the Picasso platform can be combined with various other virtualization management architectures, SDR/SDN technologies and DSA algorithms to form a fully-virtualized wireless infrastructure. The same architecture can be used for mobile handsets for more efficient antenna usage. However, according to [32], more performing and cost-effective components will be required for its implementation in commercial devices.

References

1. S. Mangold, S. Choi, G.R. Hiertz, O. Klein, B. Walke, Analysis of IEEE 802.11e for QoS Support in Wireless LANs. IEEE Wirel. Commun. **10**(6), 40–50 (2003)
2. S. Schelstraete, An introduction to 802.11ac, White paper, Quantenna communications (2011). Available (Online): http://www.quantenna.com/pdf/Intro80211ac.pdf
3. F. Fu, U.C. Kozat, in *Proceedings of IEEE INFOCOM* 2010, "*Wireless Network Virtualization as a Sequential Auction Game*," (2010)
4. Group-GENI, "Technical document on wireless virtualization," GENI GDD-06-17 (2006). Available (Online): http://groups.geni.net/geni/attachment/wiki/OldGPGDesignDocuments/GDD-06-17.pdf
5. 3rd Generation Partnership Project, "Network Sharing: Architecture and Functional Description," 3GPP TS 23.251 v11.3.0 (2012)

6. J. Sachs, S. Baucke, in *Proceedings of the 4th Annual International Conference on Wireless Internet*, "Virtual Radio: A Framework for Configurable Radio Networks," (2008)
7. R. Mahindra, G. Bhanage, G. Hadjichristofi, I. Seskar, D. Raychaudhuri, Y.Y. Zhang, in *Proceedings of 2008 Next Generation Internet Networks*, "Space Versus Time Separation for Wireless Virtualization on an Indoor Grid," (2008)
8. M. Hoffmann, M. Staufer, in *Proceedings of the IEEE International Conference on Communications Workshops*, "Network Virtualization for Future Mobile Networks: General Architecture and Applications," (2011)
9. K.-H. Kim, S.-J. Lee, P. Congdon, in *Proceedings of ACM SIGCOMM Mobile Cloud Computing (MCC) Workshop*, "On Cloud-Centric Network Architecture for Multi-Dimensional Mobility," (2012)
10. F. Boccardi, O. Aydin, U. Doetsch, T. Fahldieck, H.-P. Mayer, in *Proceedings of the 23rd International Symposium on Personal, Indoor and Mobile Radio Communications*, "User-Centric Architectures: Enabling CoMP via Hardware Virtualization," (2012)
11. G. Bhanage, I. Seskar, R. Mahindra, D. Raychaudhuri, in *Proceedings of the ACM SIGCOMM VISA Workshop*, "Virtual Basestation: Architecture for an Open Shared WiMAX Framework," (2010)
12. G. Bhanage, R. Daya, I. Seskar, D. Raychaudhuri, in *IEEE International Conference on Communications*, "VNTS: A Virtual Network Traffic Shaper for Air Time Fairness in 802.16e Systems," May 2010
13. R. Kokku, R. Mahindra, H. Zhang, S. Rangarajan, *in Proceedings of the 16th Annual International Conference on Mobile Computing and Networking*, "NVS: A Virtualization Substrate for WiMAX Networks," Sept. 2010
14. 4WARD Consortium, "Virtualisation Approach: Evaluation and Integration—Update," ICT-4WARD project FP7-ICT-2007-1-216041 Deliverable D-3.2.1, June 2010
15. Y. Zaki, L. Zhao, C. Görg, A. Timm-Giel, in *Proceedings of the Second International ICST Conference on Mobile Networks and Management*, "A Novel LTE Wireless Virtualization Framework," Sept. 2010
16. "The Click Modular Router Project," Available (Online): http://www.read.cs.ucla.edu/click/click
17. M.A. Khan, Y. Zaki, in *Proceedings of the 9th IFIP TC 6 International Conference on Wired/Wireless Internet Communications*, "Dynamic Spectrum Trade and Game-Theory Based Network Selection in LTE Virtualization Using Uniform Auctioning," June 2011
18. B. Aboba, "Virtual Access Points," IEEE 802.11-03/154rl, Mar 2003. Available (Online) https://mentor.ieee.org/802.11/dcn/03/11-03-0154-00-000i-virtual-access-points.doc
19. H. Coskun, I. Schieferdecker, Y. Al-Hazmi, *Electronic Communications of the EASST*, vol. 17, "Virtual WLAN: Going Beyond Virtual Access Points," Mar 2009
20. G. Bhanage, D. Vete, I. Seskar, D. Raychaudhuri, in *Global Telecommunications Conference*, "SplitAP: Leveraging Wireless Network Virtualization for Flexible Sharing of WLANs," Dec 2010
21. L. Xia, S. Kumar, X. Yang, P. Gopalakrishnan, Y. Liu, S. Schoenberg, X. Guo, in *Proceedings of the 7th ACM International Conference on Virtual Execution Environments*, "Virtual WiFi: Bring Virtualization from Wired to Wireless," Mar 2011
22. Y. Al-Hazmi, H. de Meer, in *Proceedings of 8th International Conference on Wireless On-Demand Network Systems and Services (WONS)*, "Virtualization of 802.11 Interfaces for Wireless Mesh Networks," Jan 2011
23. P. Serrano, P. Patras, X. Perez-Costa, B. Gloss, D. Chieng, in *Proceedings of ICT-Mobile Summit Conference*, "A MAC Layer Abstraction for Heterogeneous Carrier Grade Mesh Networks," June 2009
24. Y. He, J. Fang, J. Zhang, H. Shen, K. Tan, Y. Zhang, in *Proceedings of the ACM SIGCOMM Conference*, "MPAP: Virtualization Architecture for Heterogenous Wireless APs," Aug 2010
25. M. Bansal, J. Mehlman, S. Katti, P. Levis, in *Proceedings of the First Workshop on Hot Topics in Software Defined Networks*, "OpenRadio: A Programmable Wireless Dataplane," Aug 2012

26. K. Tan, H. Liu, J. Zhang, Y. Zhang, J. Fang, G.M. Voelker, in *Communications of the ACM*, vol. 54, "Sora: High-Performance Software Radio Using General-Purpose Multi-Core Processors," Jan 2011
27. E. Blossom, GNU radio: tools for exploring the radio frequency spectrum. Linux J. **2004**(122), 4 (2004)
28. "GNU Radio Overview," Available (Online): http://gnuradio.org/redmine/projects/gnuradio
29. D. Katabi, Technical perspective: Sora promises lasting impact. Commun ACM **54**(1), 98 (2011)
30. Texas Instruments, "TMS320TCI6616 Communications Infrastructure KeyStone SoC," Data Manual, Mar 2012. Available (Online): http://www.ti.com/lit/ds/symlink/tms320tci6616.pdf
31. K. Tan, H. Shen, J. Zhang, Y. Zhang, in *Proceedings of IEEE Symposium on New Frontiers in Dynamic Spectrum Access Network*, "Enabling Flexible Spectrum Access with Spectrum Virtualization," Oct 2012
32. S. Hong, J. Mehlman, S. Katti, in *Proceedings of the ACM SIGCOMM* 2012 *Conference on Applications, Technologies, Architectures, and Protocols for Computer Application*, "Picasso: Flexible RF and Spectrum Slicing," Aug 2012

Chapter 6
Framework for Wireless Virtualization

Since different aspects of wireless technologies are relevant for different services and applications, there are various approaches to wireless virtualization. A particular approach that is beneficial to a certain application can be detrimental to another. As there are different advantages and disadvantages of a fully virtualized system, a progressive virtualization framework that focuses on the *coexistence* of different virtualization perspectives is proposed. First, this chapter identifies three main perspectives of wireless virtualization: flow-based, protocol-based and spectrum-based virtualization. Then, it addresses the progressive integration and evolution of these perspectives in the context of a *generic wireless virtualization framework*. This is followed by the discussion of the challenges and requirements of such a framework. Finally, a *multi-dimensional* wireless virtualization framework is presented as an example architecture that can satisfy these requirements.

6.1 Perspectives in Infrastructure Virtualization

As discussed in Sect. 5.1 of Chap. 5, there are different *scopes* and *perspectives* of wireless virtualization that are suitable for different applications and services. Despite these differences, some of these architectures are inter-compatible and can complement each other's weaknesses in a fully virtualized infrastructure, as shown in Fig. 6.1. For example, from a broader infrastructure-wide or testbed-wide perspective, meta-architectures like GENI are attempts at federating different testbeds with different applications under the same framework in order to provide a unified, flexible, feature-rich and open development platform for future Internet. From the wireless infrastructure perspective, protocol-layer virtualization technologies can be combined with spectrum virtualization technologies to yield the full virtualization of the entire radio stack. Thus, different aspects of the virtualization should be jointly considered. In order to put the different architectures

H. Wen et al., *Wireless Virtualization*, SpringerBriefs in Computer Science, DOI: 10.1007/978-3-319-01291-9_6, © The Author(s) 2013

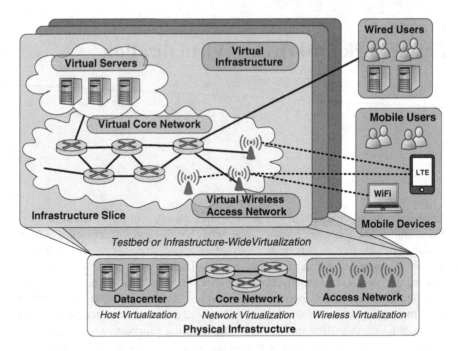

Fig. 6.1   Integration of wireless virtualization in a cloud infrastructure

and technologies surveyed in this book into perspective, their scopes and virtualization domain are shown in Table 6.1. The gray-striped cases refer to testbed architectures that do not explicitly address particular aspects but range across multiple domains.

6.1.1   Testbed, Host and Network Virtualization Domains

As shown in Fig. 6.1, a fully virtualized infrastructure is formed by the *coexistence* and *cooperation* of different *virtualization domains*. In this chapter, the term *virtualized infrastructure* refers to the virtualized computer, network and wireless hardware and equipments. The term *virtualization domain* refers to a set of components in the virtualized infrastructure sharing similar types of resources and performing similar functionalities. For example, an access network formed by wireless technologies constitutes the *wireless domain*. The virtualized version of the wireless access network is then referred to as the *wireless virtualization domain*. The term *tenant* refers to the owner of an *active* virtual instance of the infrastructure, also called a *virtual slice*. Using this terminology, the components in the same domain can form a common *resource pool* that can then be managed and allocated in a service-oriented manner. One interpretation is to view the different types of virtualization not as independent technologies but as

Table 6.1 Scope and aspects addressed by different virtualization architectures

Architecture / Project	Section / Subsection	Meta-architecture	Specific Application	Cloud Infrastructure	Operating System	NIC	Switching/Routing	Flow-based Functions	General Architecture	Service/application	Virtual BS/AP	Wireless NIC	Resource Scheduling	MAC Functions	PHY Functions	Spectrum	RF circuit
		Testbed Domain		**Host Domain**		**Network Domain**			**Wireless Domain**								
GENI	3.1		▨	▨	▨	▨	▨	▨	▨	▨	▨	▨	▨	▨	▨	▨	▨
PlanetLab	3.2	■															
SAVI	3.4	■	■	■	■								▨	▨	▨	▨	▨
VANI	3.4.1	■	■	■	■												
AKARI	3.5	■	▨	▨	▨	▨	▨	▨	▨	▨	▨	▨	▨	▨	▨	▨	▨
VNode Project	3.5.3	■						■									
OpenFlow	4.1							■									
OpenFlow Wireless	4.1.4							■									
CAPWAP	4.2.1										■		■				
CloudMAC	4.2.2			■										■			
SR-IOV + Crossbow	4.3				■												
Mobile Network Virtualization	5.3.1								■								
Configurable Radio Network	5.3.2								■				■				
Cloud-centric architecture	5.4.1			■					■								
User-centric CoMP	5.4.2														■		
vBTS	5.5.1										■						
VNTS	5.5.2												■				
NVS	5.5.3												■				
Virtual eNodeB	5.5.4										■						
SplitAP	5.6.1												■				
Virtual Wi-Fi	5.6.2				■												
PSM and PCF virtualization	5.6.3													■			
Carrier grade mesh network	5.6.4						■										
MPAP + Sora SDR	5.7.1													■	■		
OpenRadio	5.7.2													■	■		
SVL	5.8.1															■	
Picasso	5.8.2																■

a *unified ecosystem* of different domains. The main domains shown in Fig. 6.1 and Table 6.1 are identified as the *testbed domain*, the *host domain*, the *network domain* and the *wireless domain*. As shown in Table 6.1, the various architectures and implementations discussed in previous chapters are often cross-domain and can cover more than one aspects of virtualization.

The testbed virtualization domain (or *infrastructure virtualization domain*) refers to any virtualized infrastructure that supports a wide range of applications. It can contain the other three virtualization domains as part of its framework by handling the interworking among them. *Virtual infrastructure* slices are constructed over the physical infrastructure by binding resources across the other domains. Since the main focus of testbed virtualization is to provide a scalable and sustainable infrastructure for experimentation, this domain is mainly represented by the future Internet research testbeds covered in Chap. 3. The aspects covered by this domain include the federation of different testbeds and the meta-control or meta-management frameworks.

The physical components of the host virtualization domain (or *computer virtualization domain*) are the server clusters located in datacenters. This domain can also be extended to include any hardware that contains computing resources such as personal computers and mobile devices (mobile operating system virtualization). In this domain, resources such as processing cycles, access memory and storage memory can be shared among different virtual machine (VM) instances. Cloud computing is the result of these resources being *pooled* and *dynamically provisioned*. Software services such as cloud-based storage and cloud servers are examples of applications that can be supported in the host virtualization domain. This domain is important in a virtualized infrastructure since most if not all the functions in the testbed, network and wireless domain require some form of processing. Thus, many of these functions can be relocated into VM servers residing in the cloud. As such, the host virtualization domain is the processing center of the virtualized infrastructure.

The network virtualization domain is composed of core network routers and switches, as well as network interface cards (NIC). In some sense, NICs can be considered as part of the host domain since they are located on host computers. In this book, NIC is viewed as the *boundary interface* between the host and the network domain. In addition to routers and switches, the network domain can also contain specialized services and equipments located in the network infrastructure such as deep packet inspection (DPI) and firewalls modules [1]. For the virtualization of the network domain, the architectures and technologies presented in this book (Chap. 4.) are mostly *flow-based*. This flow-centric approach to network virtualization is well justified by the simple fact that one of the main functions of the network fabric is the *transport* and *redirection* of data among hosts. Thus, the abstraction and decoupling of these functionalities from the hardware through SDN is an important step in the virtualization of the network domain. In the virtualized infrastructure, the network domain is responsible for providing quality-of-service (QoS) guarantees and differentiated services to a virtual slice of the infrastructure.

Finally, the wireless domain is composed of wireless hardware equipments such as cellular basestations, wireless access points and wireless NICs. The wireless domain is usually interfaced with the network domain as an *extension* that provides *high mobility* access to the infrastructure. The wireless domain has very few applications when it is isolated from the other domains. Because of this dependence of the wireless domain on the network domain, many of the wireless virtualization architectures are considered as direct extensions to network virtualization. However, there are many aspects unique to the wireless domain, as shown in Chap. 5. The different perspectives in the wireless virtualization domain that addresses these aspects are identified in the next subsection.

6.1.2 Wireless Virtualization Perspectives

In the wireless domain, specific applications and design goals lead to different virtualization perspectives, as discussed in Chap. 5. Flow virtualization can be adequate for mobility-related applications such as the cloud-centric mobility management framework Carmen (refer to Sect. 5.4.1). However, network-layer virtualization is often not sufficiently granular for more advanced applications such as the coordinated multipoint (CoMP) using virtualization (refer to Sect. 5.4.2). Fundamentally, wireless resources should be partitioned or sliced for each tenant. In this book, the different wireless virtualization architectures and implementations can be grouped into three main perspectives based on the type of resources being virtualized and the depth of the slicing: *flow-based virtualization, protocol-based virtualization* and *spectrum-based virtualization*. The terms *full wireless virtualization* and *fully-virtualized infrastructure* refer to an architecture that includes all three perspectives. However, as the integration across the perspectives increases, the distinction between them could eventually become less important.

1. *Data* and *flow-based* **wireless virtualization**: A flow can be defined as a stream of data sharing a common signature. Flow-based virtualization is often referred to as *mobile network virtualization* because it originated from network virtualization. It is inspired by flow-based SDN technologies such as OpenFlow. Thus, it focuses on providing isolation, scheduling, management and service differentiation between both uplink and downlink traffic flows from different slices. It can be implemented as an *overlay* filter and software switch module over the existing hardware. OpenRoads (refer to Sect. 4.1.4) and vBTS (refer to Sect. 5.5.1) are examples of *overlay flow-based virtualization*. While this type of implementation can use commercial hardware as a black box component, it has limitations in terms of the granularity of the control and the resource allocation. Alternatively, virtualization can also be integrated into the internal MAC scheduler in order to use radio resource scheduling algorithms to reinforce the QoS contract of each slice. Otherwise, due to the varying nature of the wireless channel, QoS cannot be fully guaranteed in the overlay mode by traffic shaping alone. NVS (refer to Sect. 5.5.3) and the virtualized eNodeB (refer to Sect. 5.5.4) are examples of architectures that use a modified scheduler to isolate the resource scheduling of each slice. With flow-based virtualization alone, all virtual slices share the same wireless protocol stack.

2. *Protocol-based* **wireless virtualization**: Unlike the flow-based perspective, the protocol-based virtualization is a unique perspective to wireless virtualization. It focuses on the isolation, customization and management of multiple wireless protocol instances on the same radio hardware. Here, the type of resources being sliced can vary depending on how the wireless protocol is being processed on a given hardware platform. For the MAC layer, it can involve protocol decomposition as well as scheduler virtualization. For the PHY layer, it can

include the allocation of hardware DSP resources. Different degrees of protocol virtualization are possible. A *partial* implementation allows multiple instances of the same protocol stack to share the radio with different sets of configuration. Thus, different instances of MAC and PHY configuration settings can coexist on the *same* protocol stack. The existing MAC and PHY functionalities can be modified and abstracted. The PSM and PCF-based virtualization of wireless NICs (refer to Sect. 5.6.3) is an example of partial protocol-based virtualization with some enhancements to the wireless MAC. The *full* implementation of protocol-based virtualization potentially allows *different* protocol stacks to operate on the same radio hardware. This is possible using programmable SDR technologies such as Sora (refer to Sect. 5.7.1) and OpenRadio (refer to Sect. 5.7.2). These implementations can simultaneously support different wireless protocol stacks on the same radio frontend by *decoupling* the wireless protocols from the physical hardware.

3. *RF frontend* **and** *spectrum-based* **wireless virtualization**: In the protocol-based virtualization, in order for multiple protocols to coexist on the same hardware, a mechanism to share the spectrum must be applied. Otherwise, the radio platform remains merely a programmable hardware (equivalent to a SDR) but not a fully virtualized one. The third perspective, RF frontend and spectrum-based virtualization, addresses these issues. Spectrum-based virtualization and RF frontend virtualization are the deepest form of slicing currently possible. It involves the *abstraction* and *dynamic allocation* of the frequency spectrum to each tenant through spectrum reshaping and radio slicing techniques. It allows a given wireless protocol stack to use arbitrary and potentially non-contiguous frequency bands. In addition, this perspective completely decouples the RF frontend from the protocols, allowing a single frontend to be used by multiple virtual wireless nodes or multiple frontends to be used by a single node. It is implemented through cognitive radio technologies and dynamic spectrum allocation/access (DSA) techniques. In the digital domain, SVL (refer to Sect. 5.8.1) virtualizes the spectrum by using intermediate signal processing techniques to perform spectrum re-mapping, shaping and DSA. In the analog domain, Picasso (refer to Sect. 5.8.2) presents modifications to the RF frontend circuit to allow full duplex transmission of signals on the same antenna with improved isolation between spectrums of different slices.

There is some degree of overlap and synergy between these perspectives. For instance, to some extent, the integration of virtualization in the MAC scheduler of cellular basestations can be considered as a form of partial protocol virtualization (due to modifications to MAC scheduler) as well as flow-based virtualization (different scheduling for different traffic flows). Thus, these perspectives can be *complementary* to each other in a fully virtualized wireless infrastructure. Many applications cannot take full advantage of the benefits of wireless virtualization when some of these perspectives are missing. These perspectives are summarized in Table 6.2. These three main perspectives will be revisited in Sect. 6.3.2 in the context of a multi-dimensional wireless virtualization framework.

Table 6.2 Summary of different wireless virtualization perspectives

	Flow-based	Protocol-based	Spectrum-based
Resource type	• Traffic/packet flow (overlay) • Radio resource blocks (overlay + integrated)	• MAC/PHY configurations (partial + full) • MAC decision logic (full) • PHY processing blocks (full)	• Digital/analog signal spectrum • Antenna and RF frontend
Affected layers	• Transport/Network (overlay + integrated) • MAC/PHY (integrated)	• MAC/PHY	• PHY, Sub-PHY (spectrum layer) • RF frontend
Example functionalities	• Network topology • Flow isolation + provisioning • Mobility management and handover	• Power management • Coordinated multipoint (CoMP) • Wireless infrastructure-as-a-service	• Dynamic spectrum allocation/access • Full-duplex transmission • Wideband radio slicing and isolation
Example architectures or implementations (Section/ Subsection)	• OpenFlow wireless (4.1.4) • vBTS/VNTS (5.5.1, 5.5.2) • NVS (5.5.3) • Virtual eNodeB (5.5.4) • SplitAP (5.6.1) • Carrier-grade mesh network (5.6.4)	• User-centric CoMP (5.4.2) • NVS (5.5.3) • Virtual eNodeB (5.5.4) • PSM and PCF virtualization for NICs (5.6.3) • Carrier-grade mesh network (5.6.4) • MPAP + Sora (5.7.1) • OpenRadio (5.7.2)	• Picasso (5.8.1) • SVL (5.8.2)
Related concepts and technologies	• SDN	• SDR	• Cognitive radio, DSA

6.1.3 Coexistence of Different Virtualization Domains

It is interesting to note that many of the architectures and technologies surveyed in this book cover functions that are *complementary* to each other. For instance, applications such as the mobile personal grid (MPG) discussed in Sect. 5.4.1 take advantage of the *interoperation* of the different virtualization domains in order to provide a fully service-oriented infrastructure. As previously discussed, the wireless virtualization domain can be considered as an extension to the network virtualization domain. In the near future, most users will be accessing services and applications through mobile devices, making the wireless virtualization domain very important. One possible approach to allow the coexistence of different virtualization domains is the decoupling of the hardware from its functionalities in order to increase the programmability and flexibility of the infrastructure. Examples of such approach include the use of software-defined networking (SDN) and software-defined radio (SDR) technologies. By 'opening' the hardware functionalities to external control, the wireless infrastructure can be customized based on *service* and *application-oriented* principles. The Table 6.3 summarizes some examples of the benefits of wireless virtualization and the virtualization perspective required to achieve them.

The flexibility, isolation and granularity of control over the wireless infrastructure are enhanced as the depth of virtualization increases. For example, as seen in Table 6.3, flow-based virtualization perspectives are usually limited in the network management functions such as network provisioning and mobility management. Thus, a purely flow-based perspective does not address the special functions of different wireless protocols and wireless technologies. Similarly, protocol-based virtualization cannot address the fundamental limit in the RF frontend hardware. As mentioned in Sect. 5.8.3, isolation among virtual slices can be difficult if the radio frontend has poor isolation between adjacent spectrum bands. This suggests that 'deeper' virtualization perspectives might be required in order to improve the

Table 6.3 Different Benefits and Virtualization Perspectives

Potential application or benefits	Virtualization perspectives	Section/Subsection
QoS isolation between flows in different virtual networks or slices	• *Flow-based virtualization*: integration with the resource scheduler of the wireless access point or basestation	4.1 (OpenFlow), 5.5
	• *Protocol-based virtualization*: MAC or PHY protocols may be enhanced or modified to provide isolation	5.6.3 (PSM and PCF-based)
Policy management and scheduling isolation between the MVNOs	• *Flow-based virtualization*: scheduler virtualization with a flexible policy management framework	5.5.3 (NVS)
Mobility management, easier deployment and management of a dense and heterogeneous network	• *Flow-based virtualization*: infrastructure provisioning and traffic policy isolation among slices	5.4.1 (cloud-centric management)
	• *Protocol-based virtualization*: abstraction layer for MAC functions (such as handover) and SDR platforms	5.6.4 (carrier-grade mesh), 5.7 (SDRs)
Facilitate the implementation of joint PHY processing techniques such as coordinated multipoint (CoMP)	• *Protocol-based virtualization*: PHY layer virtualization and decoupling from the physical basestation	5.4.2 (user-centric CoMP), 5.7 (SDRs)
Easier integration of cognitive radio	• *Spectrum and RF frontend virtualization*: provide abstraction layer for dynamic spectrum allocation techniques	5.8

isolation between virtual slices, support more advanced functionalities and facilitate future research in wireless technologies.

Nevertheless, wide-scope virtualization perspectives cannot simply be replaced by more granular virtualization perspectives. For instance, the flow-based virtualization perspective remains a key component in nearly all virtualized wireless infrastructure. This is because protocol and spectrum virtualization still require a flow-based management framework to classify and isolate the network traffic among slices. In a sense, this suggests that data and flow-based virtualization is a *required* but not *sufficient* approach in advanced applications of wireless virtualization. On the other hand, a fully virtualized wireless hardware with highly-granular control is not necessarily suitable for mass deployment due to the complexity that it can introduce. Therefore, a meta-management framework is required in a virtualized infrastructure in order to integrate a heterogeneous mixture of *different virtualization perspectives*, adjustable based on the need of specific services and applications. This requirement of *evolvability*, not only to support emerging applications but also to support emerging infrastructure architectures as virtual instances (refer to Fig. 6.1), will be discussed in Sect. 6.2.

The decoupling of functions from hardware to software using SDN and SDR technologies brings out a very good question: *does wireless virtualization lead to a 'flattening' of layers?* As the functionalities of different layers of the traditional protocol stack become pooled into a common framework, one can argue that there are now two main layers: the *application layer* and the *infrastructure layer*. In this new paradigm, by combining the different virtualization perspectives, different wireless functionalities can be controlled and managed by a single application layer, as shown in Fig. 6.2. In this case, it does not mean that the traditional layers will disappear. Instead, they will be relocated inside new *infrastructure applications*. In other words, the virtualized infrastructure can support the legacy layered architecture as well as new types of protocols and system architectures, all residing as applications over the hardware. The decoupled architecture will also facilitate the implementation

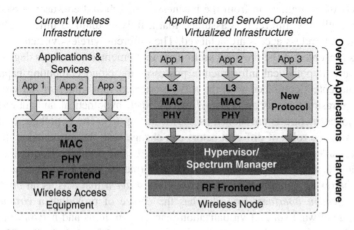

Fig. 6.2 Virtualized wireless infrastructure as a new paradigm

of functionalities that previously required joint or cross-layer integration, such as the hybrid automatic request query (HARQ) between the MAC and the PHY layer. Another notable and important application is QoS, which requires coordination across multiple domains and layers of the infrastructure. Of course, high degree of decoupling is hard to achieve. The extent of the actual implementation will vary depending on the availability of technologies and the demand of the users.

6.2 Towards a Generic Wireless Virtualization Framework

Since the different wireless virtualization perspectives are complimentary to each other, a common *generic framework* can be applied to unify these perspectives. As suggested in the previous section, the process of integration of virtualization technologies in a virtualized infrastructure should be *evolutionary* or *progressive*. At the same time, the architecture should promote the *open access* of functionalities in order to increase the flexibility of the framework and leave room for further development. This section identifies the different challenges and requirements for the design of such a generic wireless virtualization framework. First, a hypothetical virtualized ecosystem is presented as an example scenario of a fully-virtualized infrastructure. Then, the advantages and disadvantages of different virtualization architectures and implementations surveyed in this book are examined in order to identify the potential challenges of wireless virtualization. Finally, different requirements for a generic wireless virtualization framework are identified.

6.2.1 A Hypothetical Virtualized Infrastructure

In this subsection, a hypothetical scenario is projected in order to explore the potential effects of virtualization from the business, research and end-user perspectives. In this hypothetical future, it is assumed that a fully virtualized infrastructure and a ubiquitous wireless access are achieved. The different wireless virtualization perspectives presented in the previous section are implemented to various degrees.

First, this hypothetical infrastructure possesses the cloud computing capabilities of the datacenters combined with the service-oriented capabilities of the virtualized network infrastructure (as an *extended cloud*). Such hypothetical and futuristic virtualized infrastructure is envisioned as shown in Fig. 6.3. The hypothetical scenario predicts how the roles of different actors can change with the introduction of a *fully virtualized* and *integrated* wireless infrastructure. In this scenario, Eric is the equipment manufacturer. He provides the high-performance and *virtualization-enabled* wireless hardware equipment along with a *control agent* that acts as an *abstraction* and *management interface*. Eric also has the choice of providing a *virtualization platform*, a software suite of virtualization management tools and network operating systems. Otherwise, third-party companies can develop these platforms, which can access the hardware through the control agent. Irene, the infrastructure provider, owns

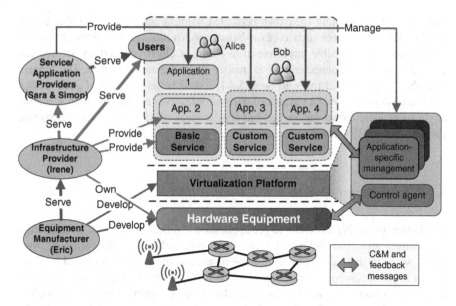

Fig. 6.3 Interaction among different entities in a hypothetical virtualized infrastructure

the hardware infrastructure. Irene can offer virtual instances of the infrastructure to the application, content and service providers Sara and Simon. Of course, Irene herself can act as a basic service provider. However, she also sells to Sara and Simon the right to provide their services over the same infrastructure without having to own it. Sara and Simon, who traditionally have less control over the delivery of their service or content through the network infrastructure, are now able to use customized network topologies and optimized protocols to accompany their services. They will be able to manage their slice via an *application-specific management* without being affected from each other. An arbitrary number of application-specific control and management (C&M) entities can be supported depending on the number of services and applications. To remain flexible and globally accessible, they can reside in the cloud. Ultimately, the users Alice and Bob are unaware of the interaction between the infrastructure and service providers. Alice and Bob simply use customized applications and services that take advantage of the fully virtualized infrastructure. Since these enhanced services are perfectly integrated within the virtualized and ubiquitous wireless access network, Alice and Bob are able to access and use them at anytime and anywhere. Finally, virtualization can allow access points bought by Alice and Bob to run virtual network applications provided by service providers, giving Sara and Simon access to the functionalities of user-bought equipment to enhance their services.

While this scenario might sound similar to the current practice of network sharing, there are important differences that can be highlighted with the help of some futuristic applications. One example application would be a power utility provider can have *full control* over the QoS scheduling of a secure relay network for smart grid communications while sharing the same wireless infrastructure with the public access network. Similar premium *virtual infrastructures* can be setup for futuristic urban services such as *smart cities*. This is currently impossible to realize without the utility providers installing their own dedicated wireless infrastructure. Other examples include new services such as *cloud-based mobility management*. Similar to cloud storage services accessible anywhere at any time, a cloud connectivity service can guarantee an always connected state no matter where the user is located. The concept of *mobile personal grid* (refer to Sect. 5.4.1) would be an example of such service that can be deployed over a virtualized wireless infrastructure. Finally, such scenarios might seem idealized or far-fetched but similar ideas and benefits for infrastructure virtualization are discussed and analysed in [2] and [3].

6.2.2 Challenges of Wireless Virtualization

In addition to having different target applications, different wireless virtualization perspectives also have different concerns and practical design difficulties to overcome. Some of these difficulties are implementation-specific, often caused by limitations in the technologies in terms of speed and complexity. Other difficulties are on a more conceptual level. For instance, the choice of partitioning and decoupling of functionalities can affect the flexibility of the implementation. Additionally, the design limitations and strict requirements of different wireless standards also present some challenges in the integration of a virtualization layer. A comparison and summary of the strong points and potential concerns of various wireless virtualization architectures and techniques surveyed in Chaps. 4 and 5 is presented in Table 6.4.

There are a few recurrent concerns and challenges common among the various virtualization architectures and implementations: the flexibility-performance trade-off, the scalability of the implementation and the added complexity of virtualization.

1. **Flexibility-performance trade-off**: Wireless technologies are inherently specialized and optimized, making it hard to apply the same virtualization approach to diverse wireless access technologies without losing their respective benefits. Thus, there is a flexibility-performance trade-off. The virtualization framework should have the flexibility of running and managing applications over diverse wireless technologies in a converged and seamless operation. However, over-generalization can lead to a reduced efficiency in the performance of a given wireless technology. Furthermore, high granularity of programmability and control of wireless resources can add overhead latency to the time-sensitive functionalities. On the other hand, a deeply integrated and specialized virtualization of a specific technology can lead to a more efficient design, albeit a less flexible and portable one. It all comes down to the question of determining the 'right amount' of wireless virtualization. Unfortunately, different applications and services have different requirements. It will be difficult if not impossible to implement a 'catch-all' framework. One potential take on the flexibility-performance trade-off is the combined use of local virtualization agents, as in VANI (refer to Sect. 3.4.1), and semi-persistent rules and policies, as in OpenFlow (refer to Sect. 4.1).

Table 6.4 Comparison of various wireless virtualization technologies

	Sects.	Strong points (Advantages)	Concerns and challenges
OpenFlow wireless	4.1.4	• Provide network flow management for MVNOs (through controller) • Integration with SDN and network virtualization	• Lack of advanced control for wireless functionalities • Scalability and fault-tolerance
CAPWAP	4.2.1	• Standardized C&M framework for potentially heterogeneous wireless networks • Decoupled functionalities from hardware	• Limited scalability of SplitMAC (controller bottleneck) • Synchronization of states between AC and WTP
CloudMAC	4.2.2	• Flexible allocation of VAP instances in the cloud	• Delay limitations for time-critical functions (i.e. 802.11 ACK)
vBTS/VNTS (WiMAX)	5.5.1 5.5.2	• *Full instance* of virtual basestation in VMs • *High portability* (with basestation equipment as black box component)	• Limited control over actual QoS and scheduling • Duplicate functionalities between virtual and physical basestation
NVS	5.5.3	• *Integrated* virtualization support *within* basestation equipment • Provides advanced and customized scheduler virtualization for MVNOs	• Increased complexity of scheduling virtualization mechanism • Does not address network-wide functions (only focused on individual basestation)
Virtual eNodeB	5.5.4	• Natural integration of virtualization in eNodeB through PRBs scheduling	• No implementation
SplitAP	5.6.1	• Handling of uplink virtualization	• Requires plugin on client station devices or user equipment (UE)
Virtual Wi-Fi	5.6.2	• Efficient virtualization of wireless NIC for VMs	• Technology-specific (802.11) • Limited support for more diverse wireless functionalities
PSM and PCF virtualization (802.11)	5.6.3	• Usage of existing MAC layer functionalities • Multiple simultaneous connections for a single NIC	• Technology-specific (802.11) • Limited scalability due to overhead • Increased complexity of channel access

(continued)

Table 6.4 (continued)

	Sects.	Strong points (Advantages)	Concerns and challenges
Carrier grade mesh network (CARMEN)	5.6.4	• Support for *heterogeneous mesh networks* • Abstraction of MAC functions as *modules*	• Some latency-sensitive functionalities are difficult to modularize
MPAP + Sora SDR	5.7.1	• Full decoupling of radio hardware from baseband processing (now performed in GPP and software) • Support for heterogeneous technologies	• Increased GPP and bus throughput requirements for higher frequency and bandwidth • Less predictable timing (limited stability)
OpenRadio	5.7.2	• Decoupling of control/ decision and datapath plane • More consistent timing capabilities	• Processing overhead due to decoupling
SVL	5.8.1	• Abstraction of dynamic spectrum management and support of cognitive radio • Flexible usage of frequency band	• Same as MPAP + Sora SDR
Picasso	5.8.2	• RF frontend circuit to support FDD in adjacent frequency bands	• Degradation of performance in high throughput situations

2. **Limited scalability**: First, when wireless functionalities are pushed to an external controller, the scalability of the virtualized network becomes dependent on the *processing capabilities* of the controller and the *link capacity* between the controller and the hardware devices. The extreme case of such processing dependence is present in Sora SDR (refer to Sect. 5.7.1), which pushes the entire baseband processing of the radio into a general-purpose processor (GPP). Then, as the virtualized wireless network is scaled and the physical controller-node distance is increased, time-sensitive functionalities can get adversely affected and start to experience longer delays. For instance, the round-trip time (RTT) between the controller and the wireless termination point (WTP) in CloudMAC (refer to Sect. 4.2.2) is limited by the strict timing requirements of the 802.11 acknowledgement (ACK) protocol. Potential methods to tackle these challenges include the use of high-throughput low-delay optical links and the efficient partitioning of functionalities.

3. **Complexity and difficulty of integration**: Depending on the depth of slicing, virtualization can add complexity to the implementation and operation of the infrastructure. By design, the virtualization layer or the hypervisor must perform the additional virtualization-related overhead functionalities of multiplexing, slice isolation and policy reinforcement. Depending on the implementation, additional function translation and overwriting might be required. For instance, a high-level

management rule has to be accurately translated into different hardware-specific control messages within strict timing restrictions. Thus, co-ordination among multiple slices can cause complex interactions among different components. Additionally, virtualization potentially requires modifications to high-performance hardware such that they can be abstracted and remotely managed. There are varying degrees of difficulty in the integration of virtualization in different wireless technologies.

In summary, many of the existing wireless virtualization architectures are either too near-sighted, with little consideration for integration within the network infrastructure, or too far-sighted, with only a few implementation details. The goal of this section is to present an *evolutionary approach* towards a generic wireless virtualization framework that can bridge the gap between these two extremes. Some requirements of such framework will be discussed in the following subsection.

6.2.3 Requirements of a Generic Wireless Virtualization Framework

The main requirements refer to both design objectives and implementation guidelines that should be followed in order to support a fully-virtualized infrastructure. These requirements consider the performance of individual virtualization modules as well as the binding framework between them. Ideally, a generic wireless virtualization framework should be able to integrate with both the network infrastructure and the individual wireless technologies. In other words, it should simultaneously offer a generic interface to attach itself to the virtualized network (*external* or *inter-domain integration*) and an open platform to support different wireless technologies and virtualization perspectives (*internal* or *intra-domain integration*). Some of the important requirements of such a framework are elaborated as follows.

1. **Generic and modular interface**: One basic requirement of a *unified* wireless virtualization framework is to have independent yet compatible and modular interfaces for different hardware devices. This requirement has the objective of providing both high flexibility and easier integration. It can be applied to different types of interfaces, ranging from internal/intra-domain interface modules to external/inter-domain framework integration. In order to satisfy this requirement, an intermediary component *localized* and *adapted* to a particular type of virtualized resources such as the *local virtualization manager* or *agent* in PlanetLab (refer to Sect. 3.2) and VANI (refer to Sect. 3.4.1) can be used as an interface and management module. Common management, networking, medium access and baseband processing functionalities among different wireless technologies and virtualization perspectives should be incorporated into a basic library of functions accessible through this local virtualization manager. Additionally, a well-defined but flexible interface allows the C&M functions of the framework to be decoupled from its virtualization mechanisms. This

enables multiple virtual instances of the control framework or even different control frameworks to act on the same virtualization layer (refer to *requirement 4: management isolation*).

2. **Technology independence of the framework**: Since the future ecosystem cannot be easily predicted, the design of the framework should not overly rely on a particular technology. While the actual implementation of the framework will be technology-specific, the architecture of the generic framework should be as technology-agnostic as possible to allow integration between different implementations of the same framework. At the same time, the components of the framework implementation should be adapted to different wireless technologies in order to operate as efficiently as possible. In other words, a distinction should be made between the *virtualization modules* (implementation-specific) and the *binding framework* that supports it (general architecture). Keeping the binding between the virtualization modules and the C&M framework independent from specific technologies leads to an easier integration of heterogeneous technologies within a virtualized infrastructure. From a research perspective, the virtualization framework can also become a tool or a platform on which new technologies can evolve.

3. **Evolvability and extensibility of the framework**: Since the wireless ecosystem is constantly evolving, a *sustainable* wireless virtualization framework is required to leave possibilities for extensions. It is not required to have all three wireless virtualization perspectives presented in this book integrated at the same time. A progressive approach adjusted to different applications is suggested. As the depth of virtualization evolves, new extension modules and plug-ins should be addable to the framework. The potential integration and federation of emerging virtualization architectures should be considered in order to avoid reinventing the wheel. The C&M layer should be scaled accordingly to support new management functionalities associated with new perspectives and new technologies. The overhead and latency incurred from the increased complexity of the framework should be mitigated in order to sustain its evolution. For instance, the delays of time-critical tasks can be mitigated to non-time-critical tasks. Otherwise, the incurred overhead and delay can counterbalance the gains achieved by virtualization.

4. **Management isolation and resource abstraction**: Tenants have different demands and service requirements for their slice of the wireless infrastructure. The framework should provide a mechanism for customizable level of control for each tenant. It should also allow different levels of *abstraction* and *transparency* adapted to the needs of each tenant. In the ideal case, each tenant should have maximum degree of freedom to apply their own C&M framework over their slice without having to worry about conflicts with other slices. *Management autonomics* is a desirable feature to allow each slice to automatically maintain service parameters through a feedback system. A maximal autonomic management could implement self-organizing, self-configuring and self-healing functionalities, although they might not be required depending on the specific application.

5. **Support for measurement and feedback**: Through the unified interface, the underlying virtualized infrastructure should provide a way to monitor the state of both the virtual instances on behalf of the MVNOs and the physical hardware itself on behalf of the infrastructure provider. These sets of feedback information can then be used for various purposes, including optimization, management autonomics, experiment measurements and commercial service metering. Similar mechanisms already exist in large-scale virtualization testbeds such as GENI (refer to Sect. 3.1) or in modern cellular technologies such as LTE.

6.3 Multi-Dimensional Virtualization Framework for Wireless Access Networks

This section presents a prototype virtualization meta-framework that is designed to support various wireless virtualization perspectives and technologies. It should be able to provide the readers an illustrative example of a framework that can satisfy many of the requirements discussed in the previous section by reusing the concepts and techniques surveyed in this book. The framework is designed to be *generic* and *multi-dimensional* in order to allow various wireless virtualization perspectives and technologies to coexist. It is divided into two separate components: the *control and management layer* (CML) and the *virtualization layer* (VL). These two layers are decoupled in order to increase the modularity of the framework. Once again, in this section, the term *tenant* refers to the owner of an *active* virtual instance of the infrastructure or a *virtual slice*. The tenants are not necessarily limited to MVNOs. They can include the application, service and content providers. The rest of this section outlines the different logical modules and their functionalities within the framework. Overall, the CML acts as a network administrator whose decisions are influenced by the tenants through service level agreements (SLAs) and by the physical resources through feedback measurement data from the VL. The VL represents the set of virtualization mechanisms reinforcing the CML decisions and the SLAs on the physical resources. The overall architecture is based on the server-client model. The framework is outlined in Fig. 6.4 and is consistent with the architectural model for Future Internet suggested in [4].

6.3.1 Control and Management Layer

Control and management (C&M) refers to the administration, coordination, metering, policing and configuration of the infrastructure by its owner. It can be used to apply network-wide functions such mobility management and load balancing. The *control and management layer* (CML) is a management framework designed to support multi-dimensional virtualization. The focus of the CML is on the

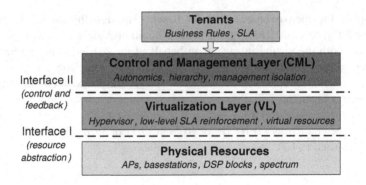

Fig. 6.4 Outline of main layers of a multi-dimensional framework

organization of the C&M functions in order to allow multiple tenants with diverse demands and business policies to co-exist on the same virtualized wireless access infrastructure. The CML is composed of two levels: the *virtual control & management layer* (vCML) and the *management hypervisor* (M-visor). The CML is outlined in Fig. 6.5. Note that the modules supported by the CML are not necessarily limited to the ones discussed in this subsection.

1. **Virtual control and management layer**: The vCML is the 'personalized' instance of the C&M framework owned by each tenant that allows them to have their own independent implementation of management functions. The vCML instances are clients to the M-visor, as shown in Fig. 6.5.

    (a) Custom management module (CMM): The CMMs contain management functions implemented and customized by the tenants. It uses the management tools and templates (MTT) as a form of API library to access the C&M of their virtual infrastructure instance. However, since not all the tenants are necessarily interested in exercising a high level of control granularity over their slice, these customized modules are optional.

    (b) Management tools & templates (MTT): MTTs are client API modules that should be called by the CMM to communicate with the corresponding *management function client* (MFC) located in the M-visor. Pre-made MFC modules should be available through the MTT API library. To address different C&M needs, three types of MTTs and corresponding MFCs are suggested in the framework, similar to the scheduler types in NVS (refer to Sect. 5.5.3). Using Type II and III modules, CMMs can be built by the tenants inside the vCML.

        1. *Type I modules* are entirely autonomous and can be configured remotely by the vCML. They are considered *out of the box* functions that tenants can use with a minimum degree of customization or knowledge on its exact operation. Type I MTTs can be directly called by the tenants without CMMs. Thus, if CMMs are not used,

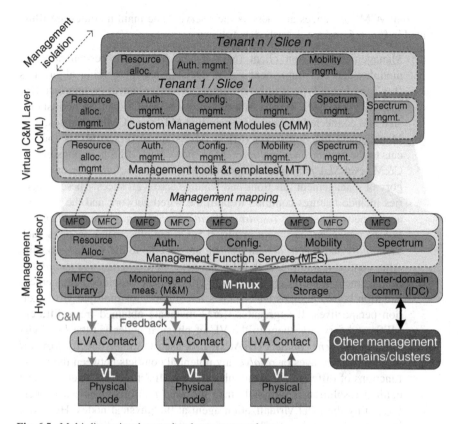

Fig. 6.5 Multi-dimensional control and management layer

the default selection of MTTs should fall back to Type I modules. For Type I modules, the entire processing is performed in the MFC.

2. *Type II modules* are semi-autonomous. They are fully-defined functions with a high degree of configurability that may actively require input from the vCML in order to operate. Unlike Type I modules, they require some basic CMMs to monitor their operation. However, most of the processing of Type II function is still performed in the MFC.

3. *Type III modules* only provide the API for low-level functions, delegating the entire logic and processing to the CMM on the vCML instances. These modules are used by tenants who want to implement their own customized functions. Thus, the processing is performed in the CMM and not the MFC. In this case, the MFC simply acts as an interface module.

2. **Management hypervisor (M-visor)**: The M-visor is an inter-slice management entity of the wireless access network composed of one or multiple *physical nodes* (access points or basestations). It provides the isolation between the

different vCML instances and acts as their server. The main modules are illustrated in Fig. 6.5 and are described as follows.

(a) Management function client (MFC): The MFCs are per-tenant agent modules that can implement C&M decision and processing algorithms (for Type I and Type II only) or simply act as a connection interface (for Type III). The *MFC library* is a repository of MFCs that can be instantiated and connected to their corresponding MTTs in vCML instances.

(b) Management function server (MFS): MFSs are servers to MFCs and clients to the *management multiplexer* (M-mux). Each MFS is responsible for C&M-level *SLA reinforcement* and C&M *conflict resolution* for one category of management functions and features across all slices. These categories include features common to both the wired network and the wireless access network, such as *resource allocation, configuration* and *authentication management*, as shown in Fig. 6.5. On the other hand, features such as *mobility* and *spectrum management* are specific to the wireless domain. After conflict resolution, each MFS can generate C&M *decision metadata*.

(c) Management multiplexer (M-mux): The M-mux provides management isolation between slices and enables the support for multiple virtualization perspectives. It aggregates C&M metadata obtained from different MFSs and forwards them to the VL of physical nodes via the *local virtualization agent contacts* (LVACs). It is used to either directly *forward* the control messages or *resolve* any potential conflicts between the C&M functions of different MFS modules (if two MFS functions overlay). The conflict resolution is optional since the final reinforcement can be performed by the local virtualization agent at the physical nodes. However, in order to maximize the degree of integration among different architectures, the M-mux can be used to perform functions that are not supported on certain physical nodes.

(d) Local virtualization agent contact (LVAC): The LVAC are clients to the *local virtualization agent* (LVA) and handle the connectivity between the M-visor and the VL of physical nodes. One LVAC is associated with a single physical node whose functions can be distributed over multiple hardware components. The LVAC acts as an abstraction layer of the physical node to hide these complexities from the perspective of the CML.

(e) Monitoring and measurement (M&M) module: The M&M module monitors the status and behavior of each physical node through feedback obtained from the VL via the corresponding LVAC. This M&M information is in turn used to support several advanced management functions in the vCML, such as management autonomics and service metering.

(f) Metadata storage module: The storage module holds information related to the status of wireless resources, such as ID number, capacity/occupancy and link health. It also stores C&M decisions and slice configuration metadata, allowing tenants to keep *configuration snapshots* of their virtual infrastructure.

(g) <u>Inter-domain communication (IDC) module</u>: The IDC module is used when a slice requires cooperation across multiple management and virtualization domains. The domains can be homogeneous/intra-domain (between wireless access domains) or heterogeneous/inter-domain (between a wireless domain and a network domain). The activity of the IDC can range from metadata synchronization to inter-domain M&M reporting and inter-domain C&M exchange.

The interconnection topology between the CML and the physical nodes is dependent on the partitioning of the functionalities and the level of centralization supported in the specific wireless access network technology. The different *management topologies* are illustrated in Fig. 6.6. In the *centralized management topology*, a single M-visor is located in a centralized server and acts as the network-wide C&M hypervisor of multiple physical nodes. Since the C&M functions are all managed through the M-visor, this mode is more suitable for technologies that have limited processing capabilities in the physical node, such as in WLANs. On the other hand, in the *distributed management topology*, an M-visor is integrated into *each* physical node and has the limited scope of a single node. In this case, network-wide functionalities are handled by the vCML instances which are directly connected to the physical nodes. Since the management hypervisor is located on the physical nodes, more powerful processing capabilities are required on the node hardware. As such, cellular basestations are more suitable for this management topology. In both cases, since vCML instances can be located on separate servers, a dedicated control channel must be established between the vCML instances and the M-visor through a potentially virtualized wired network infrastructure.

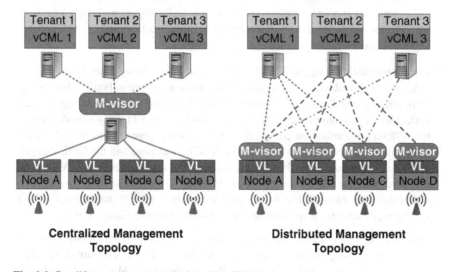

Centralized Management Topology   **Distributed Management Topology**

Fig. 6.6 Possible management topologies of the CML

6.3.2 Virtualization Layer

The *virtualization layer* (VL) is a collection of logical components that directly interact with the physical resources on the wireless equipment. Whereas the CML is a network-wide layer, the VL is local to the physical node. To maintain modularity, main logical components are separated into technology-independent *core modules* and technology-dependent *interface abstraction modules* (IAM). The *local virtualization agent* (LVA) acts as an intermediary between the CML and the VL of a physical node based on policies set through the CML. Two of the possible arrangements of the proposed VL architecture are outlined in Fig. 6.7. In the first arrangement, all virtualization core modules are located on the physical node. In the second arrangement, the flow-based modules are located on an external controller. Not all of the modules presented in this section are required in the physical nodes depending on the perspectives applied and the desired level of decoupling.

1. **Interface abstraction module**: IAMs are the binding joints of the framework designed to help satisfy the unified interface requirement. Their main goal is to keep the core modules *independent*, movable and reusable across different perspectives and implementations. Their actual function and implementation will vary depending on their location within the framework. For instance, between independent modules, they can act as inter-process communication (IPC) plug-ins.

   (a) Gateway interface module (GIM): The GIM is a special IAM that maintains *generic channels* to support control and data exchanges between the external controller and the physical nodes, as shown in Fig. 6.7. It provides a flexible channel abstraction layer that can be implemented using technologies such as VLAN tags, MPLS, GRE tunneling or OpenFlow.

   (b) Core module plug-in: The core module plug-ins are integration modules between specific wireless virtualization technologies and the LVA. They can serve as containers that map framework functionalities to technology-dependent functionalities and vice versa.

2. **Local virtualization agent**: The LVA can act either as a proxy of the CML or as an independent virtualization management system (or a mixture of both). It has a more *localized scope* and can be customized based on the hardware capabilities of each node. One of its main functions is to act as a unified control interface model among the different virtualization perspectives present on a given physical node. Each LVA has a corresponding LVA Contact (LVAC) client plugin in the CML that can setup the appropriate control and data channels. The capabilities of the LVA can be extended with the *core module plug-ins* in order to coordinate new virtualization perspectives, as shown in Fig. 6.7. Depending on the degree of integration, the LVA can manage interactions with multiple perspectives in two different modes.

   (a) Broker mode: In this mode, the LVA does not directly handle time-critical or processing-intensive tasks in order to avoid delays incurred by intermediary steps. Instead, it setups the appropriate GIM channels such that

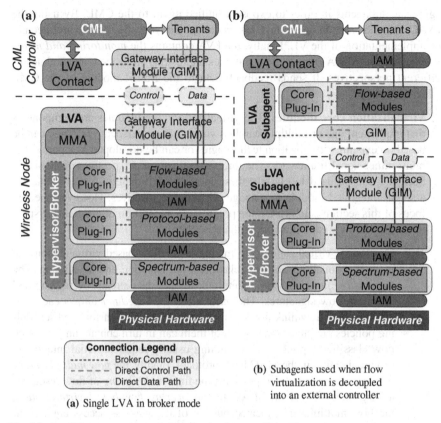

Fig. 6.7 Possible arrangements of virtualization layer modules

the tenants and the vCML instances can directly interact with the core modules within the limits determined by the SLA. The broker mode is essential in the integration of existing virtualization technologies into the VL since these technologies might have their own hypervisor and management interface.

(b) <u>Hypervisor mode</u>: In this mode, the LVA is an unified hypervisor of all perspectives and technologies supported on a given node, assuming direct control instead of setting up control channels between the modules and a third party. In this mode, the LVA can implement certain M-visor functionalities in a *distributed management topology* (refer to Sect. 6.3.1). Thus, it will be possible to directly connect the physical node to multiple vCML instances or even to another C&M framework.

In both modes, the granularity of the C&M can be controlled by adjusting the degree of direct control versus semi-persistent rules and policies. In the case when partial wireless functionalities are relocated outside of the physical node, as shown in Fig. 6.7b, the LVA can be *split* into multiple *subagents*. Then, the LVAC can aggregate subagents in order to expose a unified agent to the CML. By using the LVAC as an abstraction layer, the CML is decoupled from the specific architecture or implementation of the VL. Finally, the LVA contains the *monitoring and measurement agent* (MMA) that gathers statistics and events from the core modules of different perspectives. It sends feedback data to the M&M module of the CML when necessary.

3. **Virtualization core modules**: The virtualization core modules are components that implement one or multiple wireless virtualization perspectives. They can be based on existing virtualization technologies or can be developed as a part of the multi-dimensional framework. The three virtualization perspectives discussed in Sect. 6.1 can be used to classify these modules. Ultimately, it is possible for a single integrated module to manage all three perspectives. However, the key focus of this section is to present the progressive integration of different perspectives and how they can interwork with each other.

   (a) Flow-based virtualization core modules: For flow-based virtualization, the main functionalities are flow classification, isolation and scheduling. The *(inter-slice) flow classifier* de-multiplexes the data stream by tagging and matching flow signatures. The *intra-slice flow scheduler* allows each tenant to control downlink flows in their slice using their customized scheduling policies and algorithms. Each of them can in turn contain an intra-slice flow classifier. Uplink flow scheduling usually requires partial integration of virtualization in the MAC layer protocol. Finally, the *flow multiplexer*, or *inter-slice flow scheduler*, performs the final *multiplexing* of flow resources among slices based on SLAs. In the *overlay flow-based virtualization*, the flow multiplexer is located outside of the wireless access equipment, as shown in Fig. 6.8. In the *integrated flow-based virtualization*, all these

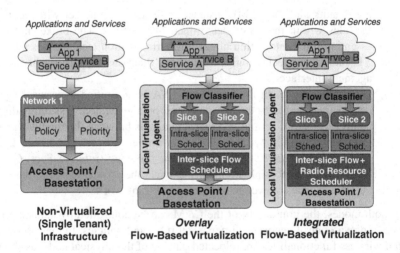

Fig. 6.8 Overlay and integrated flow-based virtualization

modules can be located inside the wireless access equipment. If protocol or spectrum virtualization is present, the flow multiplexer module might be optional since multiplexing can be performed at lower layers.

(b) Protocol-based virtualization core modules: At the protocol virtualization level, flows from different slices are directed to different *virtual wireless protocol layer* (vWPL) instances. Depending on the degree of virtualization supported, the vWPL can vary from being a simple set of protocol configurations for *partial protocol virtualization* to a fully functional virtual wireless protocol stack for *full protocol virtualization*, as shown in Fig. 6.9. In the partial protocol virtualization, a MAC or PHY functions multiplexer is used to maintain the isolation of different configuration parameters. In the full protocol virtualization where the entire protocol stack is a virtual instance, a *protocol scheduler* allocates low-level resources required for a given vWPL. These resources include access control opportunities for the MAC layer and baseband processing modules for the PHY layer. Some flow virtualization functions such as uplink flow scheduling can also be integrated within the protocol scheduler.

(c) Spectrum-based virtualization core modules: At the spectrum virtualization level, flexible portions of various frequency bands can be dynamically allocated to each slice using cognitive radio and DSA techniques. A *spectrum reshaping module* (SRM) performs intermediary signal processing to reshape and remap the baseband spectrum of each slice, as shown in Fig. 6.9. The SRM is coordinated by high-level spectrum management policies established by the CML and spectrum sensing data locally obtained.

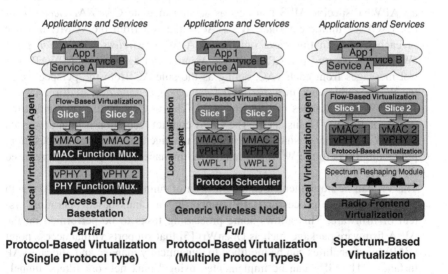

Fig. 6.9 Partial protocol, full protocol and spectrum virtualization

Techniques used in SVL (refer to Sect. 5.8.1) and Picasso (refer to Sect. 5.8.2) can be integrated in this perspective.

6.3.3 Applying the Framework in Modern Wireless Technologies

Different existing technologies and ongoing virtualization research projects can be leveraged and integrated into the implementation of the prototype multi-dimensional virtualization framework. This subsection provides examples of how the framework can be implemented in different modern wireless technologies such as the IEEE 802.11 WLAN and the LTE cellular network.

1. **Flow-based virtualization on 802.11 WLAN:** The 802.11 WLAN is widely deployed in the modern wireless ecosystem. By itself, the 802.11 standard does not define any network-wide C&M framework. It also lacks proper scheduling mechanisms for guaranteeing QoS. Thus, a virtualization framework can be applied to 802.11 WLAN to add advanced network-wide management functions and facilitate the integration of emerging applications and technologies. An example of flow-based virtualization with *broker-mode LVA* is outlined here.

    In this example, the CML plays the role of a centralized *wireless access network controller*, similar to the concept of wireless LAN controllers promoted by wireless equipment manufacturers for enterprise-class WLAN. Existing wireless network management protocols such as CAPWAP (refer to Sect. 4.2.1) can be integrated within the framework in order to support C&M functions. If the *centralized management topology* is applied, the M-visor will contain a CAPWAP-specific MFS that acts as an intermediate CAPWAP *access control* (AC) for multiple *virtual ACs* (vAC) operated from the vCML instance of each tenant. In this case, only one CAPWAP *wireless termination point* (WTP) is supported per AP. Thus, the M-mux can aggregate CAPWAP control messages from each vAC, acting as the sole AC from the perspective of the WTPs. Otherwise, in a *distributed management topology*, multiple *virtual WTP* (vWTP) instances can be supported. The vACs can directly contact the corresponding vWTP instances. The C&M conflict resolution is then processed at the WTP. However, since the distributed management topology pushes more processing tasks to the WTP, the centralized management topology is preferred for WLANs.

    The VL can be implemented as a collection of running processes on APs with a general-purpose processor (GPP) supporting a Linux-based operating system or firmware. This allows some processing tasks to be performed on the AP. A firmware package such as OpenWrt [5] that supports virtual access point (VAP) functionalities can be modified to dynamically manage and allocate VAP instances. The GIM can be implemented using Linux network filters, tunnels and OpenFlow in order to setup the data and control channels between the CML and the LVA. An OpenFlow software switch module can also be added as the intra-slice flow scheduling mechanism. In that case, an OpenFlow hypervisor can be integrated inside the CML to allow tenants to run their own controllers, similar to the case of OpenFlow Wireless (refer to Sect. 4.1.4) but with CAPWAP replacing the role of SNMP. Alternatively, OpenFlow can eventually render CAPWAP obsolete, as in the case of CloudMAC (refer to Sect. 4.2.2). Regardless of the implementation choice, the LVA acts as a common C&M interface to simultaneously manage various virtualization technologies (OpenFlow, CAPWAP, etc.).

2. **Integration of the framework in modern cellular networks**: The LTE and system architecture evolution (SAE), unlike basic 802.11, already feature a rich set of advanced management and scheduling functions that can be surprisingly well-mapped into the multi-dimensional framework. For instance, logical channels and *radio bearers* are the equivalent of flows discussed in this book. In order to guarantee the QoS in the SAE, bearers are used to identify cross-domain QoS configuration [6]. However, since the transport network is not a part of SAE, the control of the QoS over the wired network might be limited. In a sense, the integration of a generic wireless virtualization framework into the network infrastructure can solve this issue and lead to an alternative implementation of LTE over a fully-virtualized infrastructure. The virtualization of the LTE infrastructure can allow more flexible network sharing and faster evolution of the standard. In such an implementation, the CML can be integrated into the evolved packet core (EPC) in order to support new virtualization perspectives. Some of the EPC functionalities can then be consolidated into the cloud, as suggested in *network function virtualization* [1]. The CML can be used to enhance the gateway core network (GWCN) mode of network sharing in which functions like mobility management entity (MME) can be shared through a mobility MFS module. Each operator as a tenant can use a vCML to implement their own EPC. Like WLANs, the CML can be either centralized or distributed depending on where the processing functions are delegated to. The VL and LVA can be added to the eNBs in order to support virtual eNB instances and advanced spectrum virtualization techniques.

References

1. R. Guerzoni et al., Network functions virtualisation: an introduction, benefits, enablers, challenges and call for action, introductory white paper. *SDN and OpenFlow World Congress*, Oct. 2012. Available http://www.tid.es/es/Documents/NFV_White_PaperV2.pdf
2. P. Bosch, A. Duminuco, F. Pianese, and T. L. Wood, Telco Clouds and Virtual Telco: Consolidation, Convergence, and Beyond. *IEEE International Symposium on Integrated Network Management*, May 2011

3. B. Naudts, M. Kind, F.-J. Westphal, S. Verbrugge, D. Colle, and M. Pickavet, Techno-economic analysis of software defined networking as architecture for the virtualization of a mobile network. *European Workshop on Software Defined Networking*, Oct. 2012
4. A. Galis et al., Management and Service-Aware Networking Architectures (MANA) for Future Internet—Position Paper: System Functions, Capabilities and Requirements, in *Proceedings of ChinaCOM'09*, Aug. 2009
5. OpenWrt Wireless Freedom, [Online], Available https://openwrt.org/
6. H. Ekström, QoS control in the 3GPP evolved packet system. IEEE Commun. Mag. **47**, 76–83 (2009)

Authors Biography

Heming Wen received a Bachelor degree in Electrical Engineering from McGill University, Montreal, Quebec, Canada, in 2011. He is currently completing a Masters degree in Electrical Engineering in McGill University. He interned at the Canadian Space Agency, in St-Hubert, Quebec, Canada, and later at Microsoft Windows Phone Division, in Redmond, Washington, USA. His research interests include emerging wireless communications and networking topics such as wireless virtualization and next-generation wireless networks. He received the British Association Medal for outstanding academic performance at McGill University in 2012 and is currently on a masters research scholarship from the Fonds québécois de la recherche sur la nature et les technologies (FRQNT). He is part of the NSERC Strategic Network for Smart Applications on Virtual Infrastructure (SAVI).

Prabhat Kumar Tiwary completed the Bachelor of Technology in Electronics Engineering from Sardar Vallabhbhai National Institute of Technology (NIT), Surat, India, in 2011. He is pursuing his M.Eng. in Electrical Engineering at McGill University, Canada. He briefly worked as a research assistant at Wireless and Mobile Communication System Lab, Chosun University, South Korea. Currently, he is working with the NSERC Strategic Network for Smart Applications on Virtual Infrastructure (SAVI). His research interests include wireless networks, network management and wireless virtualization. He was the recipient of the "Mahatma Gandhi Scholarship" and the "Nepal Aid Fund Scholarship" in 2005 and 2007 respectively from the Ministry of External Affairs, India. He was also awarded the "Graduate Excellence Fellowship" by McGill University in 2013. Besides, he is passionate about writing fictional short stories and poems.

Tho Le-Ngoc obtained his B.Eng. (with Distinction) in Electrical Engineering in 1976, his M.Eng. in 1978 from McGill University, Montreal, and his Ph.D. in Digital Communications in 1983 from the University of Ottawa, Canada. During 1977–1982, he was with Spar Aerospace Limited and involved in the development and design of satellite communications systems. During 1982–1985, he was

H. Wen et al., *Wireless Virtualization*, SpringerBriefs in Computer Science,
DOI: 10.1007/978-3-319-01291-9, © The Author(s) 2013

an Engineering Manager of the Radio Group in the Department of Development Engineering of SRTelecom Inc., where he developed the new point-to-multipoint DA-TDMA/TDM Subscriber Radio System SR500. During 1985–2000, he was a Professor at the Department of Electrical and Computer Engineering of Concordia University. Since 2000, he has been with the Department of Electrical and Computer Engineering of McGill University. His research interest is in the area of broadband digital communications. He is a fellow of the Institute of Electrical and Electronics Engineers (IEEE), the Engineering Institute of Canada (EIC), the Canadian Academy of Engineering (CAE) and the Royal Society of Canada (RSC). He is the recipient of the 2004 Canadian Award in Telecommunications Research, and recipient of the IEEE Canada Fessenden Award 2005. He holds a Canada Research Chair (Tier I) on Broadband Access Communications, and a Bell Canada/NSERC Industrial Research Chair on Performance and Resource Management in Broadband xDSL Access Networks.